# Tributes
Volume 48

# 60 Jahre DVMLG

**Volume 38**
Logic, Intelligence and Artifices. Tributes to Tarcísio H. C. Pequeno
Jean-Yves Béziau, Francicleber Ferreira, Ana Teresa Martins and Marcelino Pequeno, eds.

**Volume 39**
Word Recognition, Morphology and Lexical Reading. Essays in Honour of Cristina Burani
Simone Sulpizio, Laura Barca, Silvia Primativo and Lisa S. Arduino, eds

**Volume 40**
Natural Arguments. A Tribute to John Woods
Dov Gabbay, Lorenzo Magnani, Woosuk Park and Ahti-Veikko Pietarinen, eds.

**Volume 41**
On Kreisel's Interests. On the Foundations of Logic and Mathematics
Paul Weingartner and Hans-Peter Leeb, eds.

**Volume 42**
Abstract Consequence and Logics. Essays in Honor of Edelcio G. de Souza
Alexandre Costa-Liete, ed.

**Volume 43**
Judgements and Truth. Essays in Honour of Jan Woleński
Andrew Schumann, ed.

**Volume 44**
A Question is More Illuminating than the Answer: A Festschrift for Paulo A.S. Veloso
Edward Hermann Haeusler, Luiz Carlos Pinheiro Dias Pereira and Jorge Petrucio Viana, eds.

**Volume 45**
Mathematical Foundations of Software Engineering. Essays in Honour of Tom Maibaum on the Occasion of his 70th Birthday and Retirement
Nazareno Aguirre, Valentin Cassano, Pablo Castro and Ramiro Demasi, eds.

**Volume 46**
Relevance Logics and other Tools for Reasoning.
Essays in Honor of J. Michael Dunn
Katalin Bimbó, ed.

**Volume 47**
Festschrift for Martin Purvis. An Information Science "Renaissance Man"
Mariusz Nowostawski and Holger Regenbrecht, eds.

**Volume 48**
60 Jahre DVMLG
Benedikt Löwe and Deniz Sarikaya, eds.

Tributes Series Editor
Dov Gabbay                    dov.gabbay@kcl.ac.uk

# 60 Jahre DVMLG

edited by

Benedikt Löwe

Deniz Sarikaya

© Individual authors and College Publications 2022. All rights reserved.

ISBN 978-1-84890-411-8

College Publications
Scientific Director: Dov Gabbay
Managing Director: Jane Spurr

http://www.collegepublications.co.uk

Cover design by Laraine Welch

---

All rights reserved. No part of this publication may be reproduced, stored in a retrieval system or transmitted in any form, or by any means, electronic, mechanical, photocopying, recording or otherwise without prior permission, in writing, from the publisher.

# Table of Contents

Vorwort
Benedikt Löwe & Deniz Sarikaya                              vii–xiii

Erinnerungen an frühe Jahre der DVMLG
Wolfgang Bibel                                                 1–11

Weihrauch complexity and the Hagen school of computable analysis
Vasco Brattka                                                 13–44

Geschichte des Lehrstuhls für Logik und Grundlagenforschung
an der Rheinischen Friedrich-Wilhelms-Universität Bonn
Elke Brendel & Rainer Stuhlmann-Laeisz                        45–50

Logic and foundations of the exact sciences at the
University of Konstanz: people & projects 1966–2021
Bernd Buldt                                                   51–96

Zum Zermelo-Ring
Heinz-Dieter Ebbinghaus & Benedikt Löwe                       97–102

DLMPS—Tarski's vision and ours
Wilfrid Hodges                                               103–119

Interview mit Arnold Oberschelp
Deborah Kant & Deniz Sarikaya                                121–126

Unterwegs mit Alan Turing
Anke Kell                                                    127–132

Mathematical logic at the Department of Mathematics
at TU Darmstadt
Ulrich Kohlenbach & Thomas Streicher                          133–137

What can formal systems do for mathematics?
A discussion through the lens of proof assistants
Angeliki Koutsoukou-Argyraki                                  139–169

Die Mitgliederentwicklung in der Frühzeit der DVMLG
Benedikt Löwe                                                 171–185

Grundlagenforschung der exakten Wissenschaften:
die DVMLG und die Philosophie
Benedikt Löwe                                                 187–202

Satzungen der DVMLG durch die Jahrzehnte
Benedikt Löwe & Deniz Sarikaya                                203–224

Eine kurze Geschichte der Entwicklung der Logik in Münster
Wolfram Pohlers                                               225–232

Logik am Mathematischen Institut der
Ludwig-Maximilians-Universität München
Wolfram Pohlers & Stan Wainer                                 233–238

The birth pangs of DLMPS
Paul van Ulsen                                                239–242

# Vorwort

Am 28. Juli 2022 jährte sich die Gründung der *Deutschen Vereinigung für mathematische Logik und für Grundlagenforschung der exakten Wissenschaften* (DVMLG) zum sechzigsten Male. Dieser Band hat zum Ziel, allen Leserinnen und Lesern einen Überblick über Hintergründe, den institutionellen Kontext und persönliche Erlebnisse von der Gründung der DVMLG bis zum heutigen Tage zu geben.

Der Band enthält historische Darstellungen von Archivmaterial, Beschreibungen von Logikstandorten im deutschsprachigen Raum und Forschungsüberblicke von Gebieten, die bedeutend für die Entwicklung des Gebiets *Logik & Grundlagenforschung* waren. Die Auswahl der Themen in diesem Bande kann keinem Anspruch der Vollständigkeit oder Repräsentativität genügen: wie immer bei Projekten dieser Art waren die Herausgeber auf die großzügige und freiwillige Zeitinvestition der Autorinnen und Autoren angewiesen, denen wir an dieser Stelle unseren herzlichen Dank aussprechen wollen.

Die DVMLG wurde zweimal gegründet: zunächst im Jahre 1954 als informelle Organisation und dann am 28. Juli 1962 als von ihrer Vorgängerinstitution separater eingetragener Verein.[1] In diesen sechzig (oder achtundsechzig) Jahren hat die DVMLG das Gebiet *Logik & Grundlagenforschung* im deutschsprachigen Raume nach innen und außen vertreten. „Nach innen" ist die DVMLG Fachvertretung unseres Gebiets im deutschsprachigen Raume; „nach außen" ist die DVMLG die Vertretung der deutschen *Logik & Grundlagenforschung* in internationalen Organisationen.

Die doppelte Gründungsgeschichte der DVMLG ist eng mit dieser Vertretung nach außen verknüpft: der Anlass für die erste Gründung im Jahre 1954 war eine Bitte des Schweizer Philosophen Ferdinand Gonseth (1890–1975) an Heinrich Scholz (1884–1956), eine deutsche Fachvertretung des

---

[1] Details finden sich in B. Löwe, *Die Mitgliederentwicklung in der Frühzeit der DVMLG* in diesem Bande, insbesondere Abschnitt 2.

| Name | Amtszeit |
|---|---|
| Arnold Schmidt (1902–1967) | 28. Juli 1962 – 16. September 1967 |
| Hans Hermes (1912–2003) | 4. April 1968 – 9. April 1970 |
| Arnold Oberschelp (geboren 1932) | 9. April 1970 – 25. November 1976 |
| Gert H. Müller (1923–2006) | 25. November 1976 – 19. September 1980 |
| Michael Richter (1938–2020) | 12. Juni 1981 – 22. April 1985 |
| Justus Diller (geboren 1936) | 22. April 1985 – 14. November 1985 |
| Klaus Potthoff (geboren 1942) | 20. Dezember 1985 – 21. Juni 1990 |
| Hans-Georg Carstens | 21. Juni 1990 – 19. Juni 1992 |
| Helmut Pfeiffer (verstorben 2004) | 19. Juni 1992 – 19. September 1996 |
| Sabine Koppelberg | 19. September 1996 – 23. September 2000 |
| Jörg Flum | 23. September 2000 – 8. August 2002 |
| Peter Koepke | 8. August 2002 – 11. September 2008 |
| *Stellv. Vors.:* Jörg Flum | 8. August 2002 – 23. September 2006 |
| *Stellv. Vors.:* Ulrich Kohlenbach | 23. September 2006 – 11. September 2008 |
| Ulrich Kohlenbach | 11. September 2008 – 13. September 2012 |
| *Stellv. Vors.:* Benedikt Löwe | 11. September 2008 – 13. September 2012 |
| Benedikt Löwe | seit 13. September 2012 |
| *Stellv. Vors.:* Volker Peckhaus | 13. September 2012 – 10. September 2016 |
| *Stellv. Vors.:* Katrin Tent | seit 10. September 2016 |

TABELLE 1. Vorsitzende der DVMLG und ihre Stellvertreterinnen und Stellvertreter von 1962 bis 2022.

Gebiets *Logik & Grundlagenforschung* für internationale Forschungsorganisationen zu bilden.[2] Die internationale Organisation, die am Ende dieser Entwicklung die globale Koordinationsrolle für das Gebiet übernahm, ist die *Division for Logic, Methodology and Philosophy of Science and Technology* (DLMPST). Wegen der besonderen Rolle, welche die Gründung dieser Organisation für die Gründung und Frühzeit der DVMLG spielte, haben wir uns entschieden, zwei Artikel zu diesem Thema in unseren Band aufzunehmen: *DLMPS–Tarski's vision and ours* ist ein Neuabdruck der Präsidialrede des DLMPS-Präsidenten Wilfrid Hodges aus dem Jahre 2011 und *The birth pangs of DLMPS* ist ein kurzer Artikel von Paul van Ulsen, der ursprünglich für die Webseite der DLMPST geschrieben wurde und in diesem Band das erste Mal in Buchform gedruckt wird.

Die zweite Gründung der DVMLG als eingetragener Verein war durch die finanziellen Konsequenzen dieser Vertretung nach außen nötig geworden: eine Vereinigung, die jährliche Mitgliedsbeiträge bezahlte, brauchte eine rechtliche Absicherung, die durch eine Eintragung ins Vereinsregister erreicht werden konnte.

---

[2]Vgl. B. Löwe, *Grundlagenforschung der exakten Wissenschaften: die DVMLG und die Philosophie*, in diesem Band, insbesondere Abschnitt 2.

| Datum | Ort | Datum | Ort |
| --- | --- | --- | --- |
| 28. Juli 1962 | Marburg/Lahn | 3. Juni 1988 | Kiel |
| 16. August 1964 | Oberwolfach | 15. Juni 1990 | Bielefeld |
| 8. April 1965 | Oberwolfach | 19. Juni 1992 | Münster |
| 9. August 1966 | Hannover | 12. Mai 1994 | Neuseddin |
| 6. April 1967 | Oberwolfach | 19. September 1996 | Jena |
| 4. April 1968 | Oberwolfach | 28. August 1998 | Berlin |
| 27. März 1969 | Oberwolfach | 23. September 2000 | Dresden |
| 9. April 1970 | Oberwolfach | 8. August 2002 | Münster |
| 1. April 1971 | Oberwolfach | 18. September 2004 | Heidelberg |
| 19. April 1972 | Oberwolfach | 23. September 2006 | Bonn |
| 12. April 1973 | Oberwolfach | 11. September 2008 | Darmstadt |
| 1. August 1974 | Kiel | 22. September 2010 | Münster |
| 13. Oktober 1976 | München | 13. September 2012 | Paderborn |
| 3. Oktober 1978 | Aachen | 4. September 2014 | Neubiberg |
| 24. August 1979 | Hannover | 10. September 2016 | Hamburg |
| 19. September 1980 | Dortmund | 15. September 2018 | Bayreuth |
| 23. September 1982 | Bayreuth | 3. März 2021 | Online |
| 18. September 1984 | Kaiserslautern | 25. August 2022 | Online |
| 16. September 1986 | Marburg | | |

TABELLE 2. Mitgliederversammlungen der DVMLG von 1962 bis 2022.

Die Geschichte der DVMLG gliedert sich grob in vier Phasen:

(i) die Vorgeschichte (1950 bis 1962),

(ii) die Frühzeit (1962 bis 1972), charakterisiert durch die jährlichen Treffen der Mitglieder in Oberwolfach,

(iii) eine Konsolidierungsphase (1972 bis 1986), charakterisiert durch eine grundlegende Satzungsänderung und die z.T. sehr hitzige Auseinandersetzung mit Grundsatzfragen und

(iv) eine Phase der Stabilität (seit 1986).

In diesem Bande finden sich einige Artikel zu verschiedenen historischen Aspekten des Vereins, darunter *Die Mitgliederentwicklung in der Frühzeit der DVMLG*, *Grundlagenforschung der exakten Wissenschaften: die DVMLG und die Philosophie* und *Satzungen der DVMLG durch die Jahrzehnte*, sowie zwei Artikel mit persönlichen Erinnerungen an die Frühzeit: Bibels *Erinnerungen an frühe Jahre der DVMLG* und das *Interview mit Arnold Oberschelp*.

In Tabelle 1 in diesem Vorwort findet sich die Liste aller Vorsitzenden der DVMLG, in Tabelle 2 eine Liste aller Mitgliederversammlungen mit Daten und Orten und in Tabelle 3 die Liste aller Vorstandsmitglieder bis zum 25. August 2022 (die an diesem Tage gewählten Vorstandsmitglieder sind noch nicht in der Liste enthalten).[3]

Die oben erwähnte Phase der Konsolidierung (1972–1986) war ein Übergang des Vereins von einer Standesvertretung mit regelmäßigen Treffen der gesamten Mitgliederschaft (im Rahmen der jährlich stattfindenden Hermes-Schütte-Tagung in Oberwolfach) zu einer breiten Fachgesellschaft, die das gesamte Gebiet *Logik & Grundlagenforschung* repräsentiert. Die Satzungen von 1962, 1967 und 1972 waren den Anforderungen einer größeren Gesellschaft nicht gewachsen (insbesondere die Tatsache, dass die Beschlussfähigkeit der Mitgliederversammlung ein Drittel aller Mitglieder erforderte) und mussten dementsprechend geändert werden; allerdings führten die sehr restriktiven Regularien zur Satzungsänderung zu großen Schwierigkeiten, diese Änderungen umzusetzen.[4]

Dieser Übergang erforderte auch eine Reflektion über das von der Gesellschaft vertretene Gebiet. In diesem Vorwort haben wir bisher den Begriff *Logik & Grundlagenforschung* als Bezeichnung des von der DVMLG vertretenen Gebiets ohne genauere Begriffsbestimmung verwendet. Der in diesem Bande enthaltene Artikel *Grundlagenforschung der exakten Wissenschaften: die DVMLG und die Philosophie* beschreibt den historischen Hintergrund und die Erwartungen der Gründer der Vereinigung, wie man diesen Begriff zu verstehen hatte, insbesondere das Verhältnis zwischen *Logik & Grundlagenforschung* und der Philosophie. In den Jahren 1978 bis 1980 entwickelte sich aus dieser Debatte ein Streit zwischen einigen Vertretern der mathematischen Logik und Vertretern der Philosophie.[5]

Der Streit um die Rolle der Philosophie in der DVMLG war harmlos im Vergleich zum Streit um das Verhältnis zwischen Logik und Informatik in den Jahren 1985 und 1986. Dieser zweite Streit führte zum Austritt

---

[3]Das genaue Wahldatum der Vorstandsmitglieder Richter und Thiel ist nicht aus den Archivdaten zu ermitteln: Die Vorstandsmitglieder Müller und Ebbinghaus schieden auf der Mitgliederversammlung in Dortmund am 19. September 1980 turnusgemäß aus, aber da diese Versammlung nicht beschlussfähig war, musste die Wahl der Nachfolger per Briefwahl erfolgen. Der Altvorsitzende Müller schrieb die Mitglieder im Oktober 1980 an und forderte sie auf, die Stimmzettel "termingemäss zurückzusenden". Es liegt weder ein Wahlprotokoll für die Vorstandswahl noch die Wahlaufforderung für die Wahl des Vorsitzenden in den Archiven der DVMLG vor, aber die neu gewählten Vorstandsmitglieder Richter und Thiel stehen bei der Wahl zum Vorsitzenden bereits zur Wahl. Die Auszählung der (später als ungültig erklärten) Wahl des Vorsitzenden hat am 12. Dezember 1980 stattgefunden (DVMLG-Archiv A93, A95, A96).

[4]Vgl. B. Löwe & D. Sarikaya, *Satzungen der DVMLG durch die Jahrzehnte*, in diesem Bande, insbesondere Abschnitt 2.

[5]Details finden sich in B. Löwe, *Grundlagenforschung der exakten Wissenschaften: die DVMLG und die Philosophie*, in diesem Bande, insbesondere Abschnitt 4.

1. Gisbert Hasenjaeger (1919–2006). 28. Juli 1962 – 9. April 1970
2. Hans Hermes (1912–2003). 28. Juli 1962 – 19. April 1972
3. H. Arnold Schmidt (1902–1967). 28. Juli 1962 – 16. September 1967
4. Kurt Schütte (1909–1998). 28. Juli 1962 – 1. April 1971.
5. Wilhelm Ackermann (1896–1962). 28. Juli 1962 – 24. Dezember 1962.
6. Paul Lorenzen (1915–1994). 28. Juli 1962 – 4. April 1968.
7. Jürgen von Kempski (1910–1998). 28. Juli 1962 – 8. April 1965.
8. Wolfgang Stegmüller (1923–1991). 8. April 1965 – 4. April 1968.
9. Arnold Oberschelp (geboren 1932). 8. April 1965 – 24. August 1979.
10. Gert H. Müller (1923–2006). 4. April 1968 – 19. September 1980
11. Ernst Specker (1920–2011). 9. April 1970 – 25. November 1976.
12. Walter Oberschelp. 1. April 1971 – 25. November 1976.
13. Heinz-Dieter Ebbinghaus (geboren 1939). 19. April 1972 – 19. September 1980
14. Justus Diller (geboren 1936). 19. April 1972 – 12. Mai 1994.
15. Helmut Pfeiffer (verstorben 2004). 25. November 1976 – 19. September 1996.
16. Anne S. Troelstra (1939–2019). 25. November 1976 – 23. September 2000.
17. Klaus Potthoff (geboren 1942). 24. August 1979 – 21. Juni 1990.
18. Michael Richter (1938–2020). November 1980 – 16. September 1986.
19. Christian Thiel (geboren 1937). November 1980 – 18. September 2004.
20. Hans-Georg Carstens. 3. Dezember 1986 – 19. Juni 1992.
21. Sabine Koppelberg. 21. Juni 1990 – 23. September 2000.
22. Martin Ziegler. 19. Juni 1992 – 28. August 1998.
23. Martin Weese. 12. Mai 1994 – 23. September 2000.
24. Wolfram Pohlers. 19. September 1996 – 11. September 2008.
25. Johann Makowsky. 28. August 1998 – 22. September 2010.
26. Peter Koepke. 23. September 2000 – 11. September 2008.
27. Wolfgang Thomas. 23. September 2000 – 23. September 2006.
28. Jörg Flum. 23. September 2000 – 23. September 2006.
29. Volker Peckhaus. 18. September 2004 – 10. September 2016.
30. Benedikt Löwe. Seit 23. September 2006.
31. Ulrich Kohlenbach. 23. September 2006 – 13. September 2012.
32. Arnold Beckmann. 11. September 2008 – 10. September 2016.
33. Ralf Schindler. 11. September 2008 – 13. September 2012.
34. Nicole Schweikardt. 22. September 2010 – 10. September 2016.
35. Sy David Friedman. 13. September 2012 – 4. September 2014.
36. Katrin Tent. Seit 13. September 2012.
37. Martin Ziegler. Seit 4. September 2014.
38. Dietrich Kuske. 10. September 2016 – 23. September 2020.
39. Olivier Roy. 10. September 2016 – 3. März 2021.
40. Heike Mildenberger. Seit 10. September 2016.
41. Matthias Aschenbrenner. Seit 3. März 2021.
42. Leon Horsten. Seit 3. März 2021.

TABELLE 3. Chronologische Liste der zweiundvierzig Vorstandsmitglieder der DVMLG bis August 2022. Die chronologische Reihenfolge ist nach Datum ihrer Wahl bzw. Reihenfolge der Nennung im Wahlprotokoll bei Wahlen am selben Tage.

der meisten Informatiker unter den Mitgliedern (darunter ein ehemaliger Vorsitzender der DVMLG) und zum Rücktritt eines Vorsitzenden.[6] Obwohl einige (aber bei weitem nicht alle) im Rahmen des Streits ausgetretenen Informatiker wieder in die DVMLG eintraten, sind auch im Jahre 2022 nicht alle beteiligten Personen der Meinung, dass der Streit damals geschlichtet wurde. Aus Rücksicht auf alle Beteiligten haben wir uns entschieden, diesen für die Geschichte und Entwicklung der DVMLG wichtigen Streit nicht zum Thema einer Analyse in diesem Bande zu machen: wir überlassen dieses Thema zukünftigen Autoren, die diese Vorgänge auf der Grundlage des Archivmaterials der DVMLG aufarbeiten können.

Der Band enthält weiterhin zwei Artikel zu spezifischen Entwicklungen des letzten Jahrzehnts: Anke Kell berichtet in *Unterwegs mit Alan Turing* über eine ungewöhnliche Aktivität der DVMLG: im *Alan Turing Year* 2012, in dem weltweit der hundertste Geburtstag von Alan Turing gefeiert wurde, organisierte die DVMLG eine Tournee des Theaterstücks *Breaking the Code* von Hugh Whitemore in Deutschland und den Niederlanden.[7] In *Zum Zermelo-Ring* berichten Heinz-Dieter Ebbinghaus und Benedikt Löwe über die Auslobung eines Preises durch die DVMLG, welcher mit dem von Ernst Zermelo geerbten Siegelring seines Großvaters zusammenhängt.

Außerdem finden sich in unserem Band fünf Beschreibungen von Logik-Standorten in Deutschland: Bonn (geschrieben von Elke Brendel und Rainer Stuhlmann-Laeisz), Darmstadt (geschrieben von Ulrich Kohlenbach und Thomas Streicher), Konstanz (geschrieben von Bernd Buldt), München (geschrieben von Wolfram Pohlers und Stan Wainer) und Münster (geschrieben von Wolfram Pohlers), sowie zwei Übersichtsartikel zu zwei spezifischen Forschungsrichtungen: die Theorie der Weihrauch-Grade (im Übersichtsartikel von Vasco Brattka) und die Verwendung von automatischen Beweisassistenten (im Interviewartikel von Angeliki Koutsoukou-Argyraki).

Wir hoffen, dass die Leserinnen und Leser dieses Bandes ebenso in die Geschichte unseres Vereins eintauchen können, wie wir es im Rahmen dieses Buchprojekts vermochten. Ganz herzlicher Dank gebührt Jane Spurr vom Verlag *College Publications*, den Mitarbeiterinnen und Mitarbeitern der Universitäts- und Landesbibliothek Münster und des Universitätsarchivs Freiburg für die Bereitstellung von Archivmaterial und des Mathematischen

---

[6]DVMLG-Archiv A125 & A126.

[7]Ursprünglich waren auch Aufführungen in Großbritannien geplant, aber die dortigen Aufführungsrechte waren bis Ende 2012 zurückgezogen worden, „embargoed owing to a professional production taking place during 2012". Der genaue Grund des Embargos ist nie bekannt geworden: es hat im Kalenderjahr 2012 keine professionelle britische Produktion von *Breaking the Code* gegeben. (Zitat aus einer persönlichen E-Mail von Judy Daish, der Agentin von Hugh Whitemore, an Benedikt Löwe v. 9. September 2011.)

Forschungsinstituts Oberwolfach für die Erlaubnis, die Bilder aus ihrem hervorragenden Bilderarchiv zu verwenden, und selbstverständlich allen Autoren und Beitragenden, ohne die dieser Band nicht existieren würde.

Cambridge & Hamburg                                    B.L.    D.S.
July 2022

# Persönliche Erinnerungen an frühe Jahre der DVMLG

## Wolfgang Bibel*

Fachbereich Informatik, Technische Universität Darmstadt, Hochschulstraße 10, 64289 Darmstadt, Deutschland
E-Mail: `bibel@gmx.net`

**Vorbemerkung**

Die Logik hat durch die Erfindung des universellen Komputers und den daraus erwachsenen vielfältigen Anwendungen eine zentrale Bedeutung in einer Reihe von wissenschaftlichen und technologischen Bereichen hinzugewonnen. Die *Deutsche Vereinigung für mathematische Logik und für Grundlagenforschung der exakten Wissenschaften* (DVMLG) trägt seit sechzig Jahren die Bezeichnung dieser Disziplin in ihrem Namen.

Der Autor hat seit einem halben Jahrhundert im Umkreis der Logik wissenschaftlich gearbeitet und ist am Beginn dieser Zeitspanne Mitglied dieses Vereins geworden. Er war als Assistent bei Kurt Schütte seit Frühjahr 1966 schon vor seiner Mitgliedschaft in zwei wichtigen Aktivitäten der DVMLG involviert, an die er sich in diesem kurzen Beitrag erinnert. Es handelt sich um das Logik-Kolloquium mit hochkarätigen internationalen Teilnehmern, das vom Verein 1966 in Hannover veranstaltet wurde, sowie um das damalige vielmonatige Ringen unter den Vorständen um eine angemessenere Vereinssatzung.

Der Text schildert im folgenden Abschnitt die Ereignisse und Hintergründe zu den mit diesen beiden Aufgaben zusammenhängenden Aktivitäten in den Jahren 1966/67 aus der damaligen Perspektive des Autors. Im darauffolgenden Abschnitt wird anhand einer—von dem in die U.S.A. ausgewanderten Logiker Richard Büchi aufgeworfenen—Frage das damals weitestgehend ausgebliebene Engagement der deutschen Logiker bei der Etablierung der Informatik erörtert, die inhaltlich ja zu einem wesentlichen Teil auf dem von der Logik errichteten wissenschaftlichen Fundament aufgebaut ist.

## 1 Hannover-Kolloquium und Satzungsänderung

Die *Deutsche Vereinigung für mathematische Logik und für Grundlagenforschung der exakten Wissenschaften* (DVMLG) wurde 1962 von den sieben deutschen Logikern Wilhelm Ackermann, Gisbert Hasenjaeger, Hans Hermes, Jürgen von Kempski, Paul Lorenzen, Arnold Schmidt und Kurt

---

*Ich danke Benedikt Löwe für die Anregung zu diesem Text sowie zur ermöglichten Einsichtnahme in vier Protokolle von Mitgliederversammlungen der 1960er Jahre und Justus Diller für seine persönlichen Anmerkungen.

Schütte gegründet und im Amtsgericht Marburg als Verein eingetragen. Alle sieben Gründungsmitglieder bildeten anfangs den Vereinsvorstand mit Arnold Schmidt (1902–1967) als dessen Vorsitzender. Da Schmidt 1950 an der Universität Marburg zum ordentlichen Professor avancierte, deuten sein Vorsitz ebenso wie die Eintragung in Marburg eindeutig darauf hin, daß von ihm in seiner neuen Position letztlich und maßgeblich die Initiative zur Gründung ausgegangen sein dürfte. Er war—zusammen mit Jürgen von Kempski (1910–1998)—auch Gründungsherausgeber der 1950 gegründeten Zeitschrift *Archiv für Mathematische Logik und Grundlagenforschung*, spielte innerhalb der deutschen Logiker damals also eine unbestritten führende Rolle.

Kurt Schütte (1909–1998) war, ebenfalls seit 1950, am Lehrstuhl von Schmidt anfangs als Wissenschaftlicher Assistent und zuletzt (formell bis 1963) als außerplanmäßiger Professor tätig [3, S. 64]. Ab Band 3 des *Archivs für Mathematische Logik und Grundlagenforschung*, also ab 1957, involvierten die Herausgeber den kurz vorher habilitierten Schütte in deren Liste „unter der Mitarbeit von". Nach seinen Berufungen als ordentlicher Professor an die Christian-Albrechts-Universität Kiel im Jahre 1963[1] und an die Ludwig-Maximilians-Universität München 1966 fungierte Schütte von Band 11 bzw. 1967 an dann als Herausgeber dieser Zeitschrift.[2] Angesichts dieser engen beruflichen Bindung an Schmidt als dessen langjähriger Assistent bzw. fachlicher Kollege am Lehrstuhl ist es verständlich, daß Schütte neben Schmidt in besonderer Weise in die Entwicklungen des Vereins von Anfang an involviert war.

Zum 1. Mai 1966 begann Schütte seine Tätigkeit als ordentlicher Professor für Mathematik an der Ludwig-Maximilians-Universität in München. Seine Mitarbeiter waren anfangs Wolfgang Bibel, Justus Diller und Klaus Vitzthum [1, S. 262]. Diller hatte bereits in Kiel als Schüttes Assistent fun-

---

[1]Nach Aussage von Justus Diller in einem privaten Telefonat vom 18. März 2021 war diese Erstberufung vor allem dem dortigen Mathematiker Friedrich Bachmann (1909–1982) zu verdanken.

[2]In [3, S. 67] wird über Schütte berichtet: „He also co-founded the journal *Archiv für Mathematische Logik und Grundlagen der Wissenschaften*". Wie aus dem hier ausführlicher beschriebenen Sachverhalt ersichtlich ist, war die damalige Aussage in zwei Punkten formal nicht ganz korrekt. Als (mutmaßlich einziger) Assistent des Gründungsherausgebers Schmidt war Schütte jedoch höchstwahrscheinlich in den Gründungsprozeß des *Archivs für Mathematische Logik und Grundlagenforschung* ebenfalls eng miteingebunden.

Die genannten Karriereschritte werden in diesem Zusammenhang deswegen ausdrücklich erwähnt, weil sie in der damaligen akademischen Welt hierzulande als unabdingbare Voraussetzungen für derartige Funktionen angesehen wurden.

In diesem Kontext ist es bemerkenswert, daß Schütte schon seit 1958 (und bis 1979), also schon neun Jahre vor seiner Herausgebertätigkeit für das *Archiv für Mathematische Logik und Grundlagenforschung*, als *Consulting Editor of the Journal of Symbolic Logic* fungierte [7, S. 49], deren internationales Ansehen weit über dem des *Archivs für Mathematische Logik und Grundlagenforschung* angesiedelt war.

giert und vorher dort bei Bachmann promoviert. Von den Mitarbeitern war in diesen Jahren nur Diller DVMLG-Mitglied und zwar ab dem 9. August 1966.[3] Zum Verständnis dieses Sachverhalts sei beiläufig die folgende Besonderheit in der damaligen Vereinssatzung kurz erläutert.

Die Aufnahme eines neuen Mitglieds erfolgte im sogenannten Modus der Zuwahl. Danach mußte zur Aufnahme eines potenziell neuen Mitglieds ein ordentliches Vereinsmitglied einen schriftlichen Antrag mit Begründung stellen. Der Kandidat mußte bereits Publikationen in den vereinsspezifischen Gebieten vorweisen können und in der Regel einen wissenschaftlichen Vortrag anläßlich einer Mitgliederversammlung gehalten haben. Die Zuwahl erfolgte dann auf der Mitgliederversammlung wiederum unter relativ strikten Bedingungen hinsichtlich der Anzahl der anwesenden Mitglieder und Vorstände sowie der abgegebenen Ja-/Nein-Stimmen. An den vom Verein abgehaltenen Tagungen konnten in der Regel nur Mitglieder teilnehmen.[4] Das Verfahren zur möglichen Teilnahme weiterer Personen wurde auf der vorangehenden Mitgliederversammlung jeweils eigens beschlossen. Die DVMLG war damals insoweit ein gewissermaßen elitärer Verein.

Nur Diller konnte diese strikten Voraussetzungen zur Aufnahme eines neuen Mitglieds erfüllen, während für die wissenschaftlichen Anfänger Vitzthum und Bibel die Mitgliedschaft zu diesem Zeitpunkt noch ausgeschlossen war. Deshalb besprach Schütte Vereinsangelegenheiten vorzugsweise mit Diller. Angesichts der kleinen und täglich eng verbundenen Arbeitsgruppe war es jedoch unumgänglich, daß auch der Autor an vielen dieser Gespräche beteiligt und daher über die Entwicklung des Vereins ab 1966 gut informiert war. Meine Erinnerungen daran bilden die Grundlage der folgenden Schilderungen.

Zum besseren Verständnis des Nachfolgenden erscheint vorweg eine kurze Charakterisierung der beiden Persönlichkeiten Schmidt und Schütte und deren Verhältnis zueinander hilfreich. Obwohl ich Schmidt bei den beiden DVMLG Veranstaltungen in Hannover 1966 und Oberwolfach 1967 begegnet sein müßte, habe ich leider keine irgendwie geartete persönliche Erinnerung an ihn—ganz im Gegensatz zu vielen anderen Logikern von damals. Offenbar hat er mich nicht besonders beeindruckt. Als erfolgreicher Initiator hat er sich in jedem Fall durchaus bleibende Verdienste erworben. Aus diesen seinen herausragenden organisatorischen Aktivitäten ist aber zu schließen, daß er charakterlich eher zu einer gern führenden und, wie wir im weiteren Verlauf noch sehen werden, durchaus auch dominanten Rolle neigte.

---

[3]Protokoll der Mitgliederversammlung vom 9. August 1966, DVMLG-Archiv A41, Tagesordnungspunkt 1.a.

[4]Dieser Beschreibung der Zuwahl liegt die Satzung in der geänderten Fassung vom 6. April 1967 zugrunde, die dem Protokoll der Mitgliederversammlung vom gleichen Datum beiliegt (DVMLG-Archiv A43). Vgl. B. Löwe, D. Sarikaya, *Satzungen der DVMLG durch die Jahrzehnte*, in diesem Bande.

Von seinen wissenschaftlichen Beiträgen ist mir wenig bekannt. Vor allem in der Einschätzung von Schütte, die ich aus dessen beiläufigen Äußerungen erschließen konnte, waren sie nicht allzu bemerkenswert. Sein mit fast 600 Seiten recht umfangreiches Lehrbuch allein über die Aussagenlogik [9] ist mir nie in die Hand gekommen. Von Schütte wurde dieses vor allem auch deswegen kritisiert, weil es trotz seines erheblichen Umfangs über diese logische Grundstufe nicht hinausgekommen war.[5] Der ursprünglich geplante zweite Band ist dann nie mehr verwirklicht worden.

Schütte war charakterlich dagegen eine bescheidene und zurückhaltende Persönlichkeit, die zudem von den außerordentlich schwierigen Jahrzehnten seiner beruflichen Karriere geprägt worden war. Das Streben nach Macht und Einfluß war ihm fremd. Als Wissenschaftler allerdings ließ er, auch bei sich selbst, nur hohe Qualität und Perfektion gelten. Seiner eigenen Fähigkeiten war er sich dabei sehr bewußt und ließ sich auf seinem Weg auch nicht von anderen beirren. Sein hohes Ansehen erwarb er sich durch sein vorbildliches Verhalten, vor allem durch seine wissenschaftlichen Leistungen bis ins hohe Alter. Fachlich beschränkte er sich auf seinen wissenschaftlichen Kompetenzbereich, der immer sehr fokussiert und im Gefolge vergleichsweise eng geblieben ist. Auch seine kritischen Urteile beschränkte er auf diesen Bereich; solche über darüber hinaus gehende Bereiche oder über Personen ließ er nur im privaten und vertraulichen Gespräch verlauten.

Schmidt und Schütte waren demnach höchst unterschiedliche Persönlichkeiten und Charaktere, was sich im Verlauf der Jahre vor allem für Schütte zu einer großen Belastung auswuchs. Über die Zeit von Schütte am Lehrstuhl von Schmidt habe ich bereits in dem Beitrag [3, S. 64] einiges berichtet. Ganz offensichtlich war es für den nur sieben Jahre jüngeren Schütte alles andere als leicht, einem beherrschenden Schmidt als Untergebener mehr als ein Jahrzehnt zu assistieren. Denn Schütte erwarb sich in dieser für ihn wahrhaft schwierigen Zeit rasch ein hohes internationales Ansehen als exzellenter Logiker. Dem hatte Schmidt fachlich wenig entgegenzusetzen—mit Ausnahme seiner machtvollen akademischen Position als einflußreicher Ordinarius, was er Schütte als seinem Untergebenem über die Jahre hin deutlich spüren ließ, obwohl—oder vielleicht gerade weil—dieser seinem Chef wissenschaftlich haushoch überlegen war.[6] Das deutsche Hochschulsystem der damaligen Jahre unterstützte diese rein hierarchisch festgefügte Konstellation auf eine bedrückende Weise.

Spätestens 1966 hatte sich das Pendel aber nun auch positionsmäßig zugunsten von Schütte gedreht. Da es ihm vor allem um das Wohlerge-

---

[5] Diese Erinnerung an Schüttes Kritik daran wurde mir von Diller in dem bereits erwähnten Telefonat bestätigt.

[6] Diller charakterisierte in dem Telefonat vom 18. März 2021 das resultierende Verhältnis von Schütte und Schmidt als „sehr problematisch".

hen des Faches ging, sah er sich nun in der Lage für dieses notfalls auch gegen den offenbar verkrusteten Willen von Schmidt Verbesserungen zu erkämpfen. Der Umfang des von der DVMLG vertretenen Faches war in deren Namen zum Ausdruck gebracht. Eine genauere Interpretation der damit beschriebenen Inhalte war den Vorständen und Mitgliedern überlassen. In dieser Hinsicht gab es zwischen Schmidt und Schütte aber wohl weitgehende Übereinstimmung, denn beide entstammten dem gleichen Strang der Hilbertschen Schule, dem sie zeitlebens eng verbunden blieben.

Konfliktpotenzial ergab sich zwischen den beiden vielmehr in der Zielsetzung einer Fortentwicklung des Faches und damit auch des Vereins in personeller und organisatorischer Hinsicht. Hierzu strebte Schütte ohne jegliche eigene Ambitionen eine Änderung an der Vereinsspitze und der Vereinssatzung an. In diesem Bestreben wurde er von einer Reihe von vor allem jüngeren Kollegen unterstützt. Nur Schmidt war offenbar höchst widerwillig und deshalb nur auf erheblichen Nachdruck dazu bereit. So findet sich am Ende des Protokolls der Mitgliederversammlung vom 9. August 1966 (DVMLG-Archiv A41) der vielsagende Satz: „Der Vorsitzende sichert zu, die Satzung der DVMLG den Mitgliedern im Verlauf der nächsten drei Wochen zuzuschicken." Offenbar hatte Schmidt den Text der Vereinssatzung auch vier Jahre nach der Vereinsgründung den Mitgliedern noch nicht zur Kenntnis gebracht und quasi unter Verschluß gehalten. Erst die Mitgliederversammlung samt Protokoll mußten ihn zur Offenlegung in dieser Weise zwingen. Daraus ist zudem ersichtlich, daß Bestrebungen einer Satzungsmodifikation zu diesem Zeitpunkt bereits in vollem Gange waren.

Unter einer Fortentwicklung von Fach und Verein verstanden Schütte und seine Mitstreiter auch eine Öffnung gegenüber der internationalen Logikergemeinschaft. So liest man im Protokoll der Mitgliederversammlung vom 8. April 1965 (DVMLG-Archiv A38) unter Tagesordnungspunkt 8 den Satz: „Dem Plan eines deutsch-englischen Logikertreffens 1966 wird zugestimmt." Daß die Initiative zu einem solchen Treffen vor allem auf den durch seine vorherigen Tätigkeiten im Ausland [3, S. 64f] international bestens vernetzten Schütte zurückgeht, erschließt sich unzweifelhaft aus seiner herausragenden Rolle in der Verwirklichung dieses Plans. Dessen Ergebnis bestand in dem *Kolloquium über Logik und Grundlagen der Mathematik*, das vom 8. bis 12. August 1966 in Hannover stattfand. Die örtliche Tagungsleitung wurde Ernst Thiele anvertraut. Das Vorwort zum 1968 erschienenen Tagungsband [10] stammt infolge des Ablebens von Schmidt 1967 allein aus der Feder von Schütte.

Diese beiden Themen, Hannover-Kolloquium und Satzungsänderung, dominierten die Diskussionen in unserer kleinen Gruppe um Schütte in den Monaten ab Mai 1966. Selbstverständlich wurden wir alle in die vorbereitenden Arbeiten für das Hannover-Kolloquium voll miteinbezogen. Schütte war zusammen mit Schmidt Programmvorsitzender. So landeten viele (oder alle) eingereichten Beiträge an Schüttes Lehrstuhl und wurden zu einem beträchtlichen Teil von uns logistisch bearbeitet, teilweise auch inhaltlich besprochen. Auf diese Weise waren wir in gewisser Weise in den Auswahlprozeß miteingebunden, angesichts der mangelnden Expertise von Vitzthum und Bibel diese beiden inhaltlich natürlich nur am Rande. Gleichwohl erinnere ich mich noch an so manche sehr kritische Bemerkungen von Schütte über international durchaus bekannte, ihm thematisch aber fernerstehende Kollegen. Für mich waren diese Diskussionen sehr lehrreich und höchst interessant.

Derart intensiv vorbereitet war für mich als Doktorand dann das Kolloquium in Hannover ein wahrhaft beeindruckendes Erlebnis, an das ich mich bis heute gerne erinnere. Ein spontaner Diskussionsbeitrag von Alfred Tarski (1901–1983) im Stile einer aus dem Stegreif gehaltenen Lehrstunde am Ende des Vortrags einer Adeligen aus Rom über vermeintliche Fehler in Gödels Beweis seines zweiten Unvollständigkeitssatzes war darunter einer der für mich denkwürdigen Höhepunkte. Die Erwähnung des Namens Tarski deutet dabei auch an, welch hochkarätige Teilnehmer in diesem Kolloquium versammelt waren, das zu den herausragendsten Höhepunkten in der Geschichte der DVMLG gehört.[7]

Weit weniger erfreulich, weil aus meiner Sicht kleinkariert und insoweit für mich uninteressant, gestalteten sich für mich als quasi unbeteiligten Zuhörer die endlosen Erörterungen über die Satzung der DVMLG, die sich bis April 1967 hinzogen. Offenbar stand Schütte hierüber im Kontakt mit einer Reihe seiner Kollegen, vor allem den Vorstandskollegen. Immer wenn sich daraus neue Gesichtspunkte oder Formulierungsvorschläge ergaben, wandte er sich damit umgehend an Diller, dessen Meinungen hierzu er ganz offensichtlich sehr schätzte. War die Diskussion so einmal angestoßen, setzte sie sich auch bei weiteren Gesprächen, beispielsweise beim gemeinsamen Mittagessen in einem der nahegelegenen Wirtshäuser unaufhaltsam fort. Dabei wäre es mir viel lieber gewesen, er hätte bei diesen zwanglosen Gelegenheiten stattdessen mehr aus seinem Leben erzählt oder über seine Arbeit gesprochen. Es verwundert deshalb nicht, daß sich mein Gedächtnis weigerte, inhaltliche Details dieser Satzungsdiskussion aufzubewahren.

Unauslöschlich bewahrt hat es jedoch, daß über viele Monate mit Schmidt um eine flexiblere Satzung gerungen wurde, bis er einem für alle wenigsten einigermaßen akzeptablen Text endlich zustimmte. Diese geänderte Fassung wurde dann anläßlich der Mitgliederversammlung vom 6. April 1967 unter Tagesordnungspunkt 3 einstimmig angenommen (DVMLG-Archiv A43).

---

[7]Da ich infolge meiner späteren Laufbahn über die weiteren Veranstaltungen der DVMLG so gut wie nicht informiert bin, kann ich nicht beurteilen, ob die DVMLG nach 1966 überhaupt noch einmal eine derart international bedeutende Veranstaltung organisieren konnte.

Schmidt starb fünf Monate danach, auf welche Nachricht hin Diller Schütte gegenüber der vielsagende Stoßseufzer entfuhr: „Wieviel Mühe hätten wir uns sparen können, hätten wir sein baldiges Ende vorausgesehen!"

## 2   Büchis verwunderte Frage

In den dreihundert Jahren seit Gottfried Wilhelm Leibniz (1646–1716) bis etwa zur Mitte des letzten Jahrhunderts durfte die Logik vor allem in Deutschland eine große Blüte erleben [8]. Nicht zuletzt mit den seither in der Logik erarbeiteten Kenntnissen wurden die Grundlagen für die revolutionären wissenschaftlichen und technologischen Entwicklungen seit der Mitte des letzten Jahrhunderts gelegt. Zu diesen Entwicklungen gehören in erster Linie die Erfindung des modernen Komputers u. a. durch Konrad Zuse (1910–1995) und die darauf aufbauende Informationstechnologie (IT). Diese bis heute andauernden und sich noch immer beschleunigenden Fortschritte erscheinen ohne den hohen und vielgestaltigen wissenschaftlichen Kenntnisstand in der Logik als kaum denkbar (siehe zB. [5]).

In Kenntnis dieses unbestrittenen Sachverhalts stellte mir Richard Büchi[8] (1924–1984) in einem persönlichen Gespräch, zu dem wir 1969 in München Gelegenheit hatten [1, S. 288f], mit großer Verwunderung die folgende Frage: „Warum waren es nicht die deutschen Logiker, die den Aufbau der Informatik in Deutschland in die Hand genommen haben?"

Stattdessen wurde dieser Aufbau weitestgehend angewandten Mathematikern oder Ingenieuren überlassen, die in aller Regel von diesen entscheidenden logischen Grundlagen so gut wie keine Ahnung hatten. Büchis Verwunderung ist nicht nur fachlich sondern auch deswegen sehr verständlich, weil im Gegensatz zu Deutschland beispielsweise in den USA die dortigen Logiker einen beträchtlichen Einfluß auf den Aufbau der dortigen *Computer Science* (CS) nahmen.[9] Es sei Historikern überlassen zu analysieren, inwieweit dieser Einfluß für den beispielsweise in den USA viel erfolgreicheren Aufbau dieser neuen Disziplin mitentscheidend war.

Am mangelnden Kontakt und Informationsaustausch unter den Beteiligten kann dieser fehlende Einfluß in Deutschland nicht gelegen haben. So widmet Konrad Zuse (1910–1995) in seinen Memoiren [12, S. 99] einen eigenen

---

[8]Wie de facto Schütte [3, S. 63] (und mutmaßlich auch Schmidt) war auch Büchi Schüler von Paul Bernays (1888–1977), worauf sich eine lebenslange Verbundenheit unter allen vieren gründete.

[9]Es würde viel zu weit führen, im vorliegenden Kontext hierauf näher einzugehen. Beiläufig sei nur erwähnt, daß nicht zuletzt Büchi selbst und sein erster, 1967 bei ihm promovierter, exzellenter und in der CS dann höchst erfolgreicher Schüler Lawrence H. Landweber an diesem Einfluß substanziell beteiligt waren. Ich selbst bin in der internationalen Szene unzähligen Kollegen begegnet, die von Haus aus Logiker waren, danach aber die Entwicklung der CS in Ihren Ländern entscheidend mitprägten.

Abschnitt mit der Überschrift „Ein anregender Besuch" der persönlichen Begegnung mit dem einflußreichen Logiker Heinrich Scholz (1884–1956) im Jahre 1944.[10] Da der DVMLG-Mitbegründer Hans Hermes (1912–2003) bei Scholz 1938 promovierte, mit ihm auch nach der Promotion in engstem Kontakt blieb und 1953 sein Nachfolger in Münster wurde, dürfte zumindest über ihn die Kunde vom Logik-basierten Komputer bis zu den späteren Vorständen der DVMLG schon in frühen Jahren vorgedrungen sein. Analoges gilt für den DVMLG-Mitbegründer Gisbert Hasenjaeger (1919–2006), der 1950—ebenfalls bei Scholz—promovierte und 1962 zum ordentlichen Professor an der Universität Bonn avancierte. Er baute sogar selbst logistische Maschinen, teilweise zusammen mit seinem Schüler Dieter Rödding (1937–1984), der 1966 wiederum Nachfolger von Hermes in Münster wurde. Desgleichen hat Hermes selbst sehr früh über Rechenmaschinen publiziert [6] und beispielsweise auf der Fachtagung über Lernende Automaten am 13. April 1961 in Karlsruhe einen Vortrag über *„Die Rolle der Wahrscheinlichkeit beim Lernprozeß"* gehalten.

Alle DVMLG-Mitbegründer dürften zudem auch Zuses Publikationen—beispielsweise die im Archiv für Mathematik [11]—zumindest zur Kenntnis genommen haben, die sich explizit auf logische Themen bezogen.[11] Weiter nahmen—erkennbar für alle Teilnehmer—am Hannover-Kolloquium von 1966 Wissenschaftler teil, die sich intensiv mit Anwendungen von Komputern beschäftigten. Schließlich lagen die zentralen Fragen der Informatik—beispielsweise die der Berechenbarkeit—jedenfalls bei Logikern wie Gödel, Hilbert und vielen anderen aus dem deutschen Sprachraum schon viel früher mit im Zentrum von deren wissenschaftlichen Arbeiten. Auch gehörte die Informatik eindeutig zu den „exakten Wissenschaften", zu deren Grundlagen sich die DVMLG mit ihrem Namen ausdrücklich verpflichtete. Kurz, den deutschen Logikern mußte damals die Relevanz der Logik und damit ihre Mitverantwortung für das aufkeimende neue Fach voll bewußt gewesen sein.

Ein eingehenderes Interesse der deutschen Logiker an der dann 1970 in Deutschland formal begründeten Informatik hat sich aus all dem während der Gründungsphase leider überhaupt nicht ergeben. So findet sich unter den Vortragenden in Oberwolfach 1967 nicht ein einziger mit einer Nähe

---

[10] Zu Scholz' Rolle in der Vorgeschichte der DVMLG, vgl. B. Löwe, *Grundlagenforschung der exakten Wissenschaften: die DVMLG und die Philosophie* in diesem Bande, darin insbesondere Abschnitt 1.

[11] Eine ausführliche Beschreibung von Zuses höchst erstaunlichen Arbeiten ua. zur maschinellen Generierung von Ableitungen in einem Kalkül der Prädikatenlogik bereits in den 1940er Jahren findet sich in dem Kapitel [4, S. 730f].

Zum Verständnis der hier getroffenen Aussage muß man sich vor Augen halten, daß zu jener Zeit die Zahl der Veröffentlichungen vergleichsweise noch relativ klein, für einen Mathematiker bzw. Logiker eine Publikation im Archiv der Mathematik daher praktisch nicht zu übersehen war.

zur Informatik. Der 1968 als DVMLG-Mitglied aufgenommene[12] Heinz Gumin (1928–2008), der 1954 in Münster promovierte, später Vorstandsmitglied bei Siemens und Präsident der Gesellschaft für Informatik (GI) wurde, konnte mit seinem Beispiel an diesem im Umfeld der DMVLG mangelnden Interesse am Aufbau der Informatik ebenfalls nichts zum Positiven hin bewirken. In den Gremien zur Etablierung der Informatik fand sich denn auch nicht ein einziger Logiker [2]. Die Reaktion auf meinen Vortrag über „*Ein mechanisches Beweisverfahren für die Prädikatenlogik*" auf der DVMLG-Tagung in Oberwolfach im April 1972 war ebenfalls von spürbarem—und von einem jungen Wissenschaftler schmerzlich empfundenem—Desinteresse geprägt, obwohl es darin ja eigentlich um ein genuines Thema der Logik ging.

Aus all dem und vielem mehr geht hervor, daß Büchis Frage in vielfacher Hinsicht sehr berechtigt war. Was aber könnte die Antwort darauf sein?

Es muß Historikern der Versuch überlassen bleiben, eine solche Antwort in schlüssiger Weise zu formulieren. Ich kann hier nur auf einen Aspekt hinweisen, der nach meiner Einschätzung sicher eine gewisse Rolle gespielt hatte. Er läßt sich sogar bis zu einem bestimmten Grade an der im letzten Abschnitt beschriebenen Satzungsdiskussion erläutern. Dort nannte ich die damalige DVMLG „elitär", weil der Zuwachs und die Beteiligung nur unter strengster Kontrolle der ursprünglichen Vereinsgründer stattfinden konnte. Eine solche Organisationsstruktur neigt in inhärenter Weise dazu, sich gegenüber Neuerungen oder Veränderungen abzuschotten.

Jede Organisation mit einer bestimmten Zielvorgabe steht vor diesem Dilemma, wie die Entscheidungsprozesse organisiert werden sollen. Läßt man dabei zu viele, vor allem auch viele inkompetente Akteure mitentscheiden, steigt mit deren Anzahl die Wahrscheinlichkeit der Zielverfehlung. Beschränkt man die Gruppe der Entscheider dagegen auf einen zu kleinen Kreis ohne die Möglichkeit der Einwirkung von ggf. durchaus kompetenten Außenstehenden, dann bleibt der wünschenswerte Fortschritt, beispielsweise in Bezug auf die Anpassung der Zielvorgabe an ein sich veränderndes Umfeld, auf der Strecke.[13]

Die Gratwanderung zwischen diesen beiden Abgründen ist eine hohe Kunst, die den Vorständen der DVMLG in den 1960er Jahren nach meiner

---

[12] Protokoll der Mitgliederversammlung vom 4. April 1968 (DVMLG-Archiv A48).

[13] Die mittelalterlichen Zünfte waren für dieses Dilemma oft ein gutes Beispiel. Richard Wagner hat es in der Oper *Die Meistersinger von Nürnberg* in unvergeßlicher Weise am Beispiel der Meistersingerzunft dargestellt. In ihr wurden danach strenge Gesangs- und Textregeln tradiert, die dem begabten Sänger Ritter Walther von Stolzing aber nicht geläufig waren, sodaß er sich daran auch nicht halten konnte und wollte. Ungeachtet vieler Regelfehler werden die Zuhörer von seinem Liedvortrag aber tief bewegt. Unter dem Druck des begeisterten Volkes nimmt dann alles doch einen guten Ausgang. Stolzing wird der Preis des Wettbewerbs zugesprochen und in die Zunft aufgenommen. Die tradierte Zielvorgabe wurde so spontan modifiziert.

persönlichen Einschätzung besser hätte gelingen können.[14] Aber die Geschichte kann man bekanntlich nicht zurückdrehen; man kann aus ihr nur für die Zukunft lernen.

## Literaturverzeichnis

[1] L. W. Bibel. *Reflexionen vor Reflexen—Memoiren eines Forschers.* Cuvillier Verlag, 2017.

[2] W. Bibel. On the development of AI in Germany. *Künstliche Intelligenz*, 34(2):251–258, 2020.

[3] W. Bibel. Reminiscences of Kurt Schütte. In R. Kahle & M. Rathjen, Hrsgg. *The Legacy of Kurt Schütte*. Springer, 2020: S. 63–69.

[4] W. Bibel. Laßt hundert Blumen blühen. In R. H. Reussner, A. Koziolek & R. Heinrich, Hrsgg. *50. Jahrestagung der Gesellschaft für Informatik, INFORMATIK 2020, Back to the Future, Karlsruhe, Germany, 28. September–2. Oktober 2020*. Lecture Notes in Informatics P-307, Gesellschaft für Informatik, 2021: S. 729–746.

[5] M. Davis. *Engines of Logic—Mathematicians and the Origin of the Computer*. Norton, New York, 2000.

[6] H. Hermes. Die Universalität programmgesteuerter Rechenmaschinen. *Mathematisch-Physikalische Semesterberichte*, 4:42–53, 1954.

[7] R. Kahle & M. Rathjen, Hrsgg. *The Legacy of Kurt Schütte*. Springer, Cham, 2020.

[8] W. Kneale and M. Kneale. *The Development of Logic*. Clarendon Press, 1984.

[9] H. A. Schmidt. *Mathematische Gesetze der Logik—I Vorlesungen über Aussagenlogik*. Grundlehren der mathematischen Wissenschaften 69. Springer, 1959.

[10] H. A. Schmidt, K. Schütte & H.-J. Thiele, Hrsgg. *Contributions to mathematical logic, Proceedings of the Logic Colloquium, Hannover 1966*, Studies in logic and the foundations of mathematics 50, North-Holland Publishing Company, 1968.

---

[14] Für mich selbst war dies aber in mehrfacher Hinsicht eine sehr lehrreiche Erfahrung für meine weitere wissenschaftliche Karriere.

[11] K. Zuse. Über den Plankalkül als Mittel zur Formulierung schematisch kombinativer Aufgaben. *Archiv der Mathematik*, 1(6):441–449, 1948.

[12] K. Zuse. *Der Computer mein Lebenswerk*. Verlag Moderne Industrie, 1970.

# Weihrauch complexity and the Hagen school of computable analysis

## Vasco Brattka*

Institut für Theoretische Informatik, Mathematik und Operations Research, Fakultät für Informatik, Universität der Bundeswehr München, Werner-Heisenberg-Weg 39, 85577 Neubiberg, Germany

Laboratory of Discrete Mathematics & Theoretical Computer Science, Department of Mathematics & Applied Mathematics, University of Cape Town, Private Bag X3, Rondebosch 7701, South Africa

E-mail: Vasco.Brattka@cca-net.de

### Abstract

Weihrauch complexity is now an established and active part of mathematical logic. It can be seen as a computability-theoretic approach to classifying the uniform computational content of mathematical problems. This theory has become an important interface between more proof-theoretic and more computability-theoretic studies in the realm of reverse mathematics. Here we present a historical account of the early developments of Weihrauch complexity by the Hagen school of computable analysis that started more than thirty years ago, and we indicate how this has influenced, informed, and anticipated more recent developments of the subject.

## 1 Computable analysis in Hagen

The Hagen school of computable analysis was founded by Klaus Weihrauch in the 1980s. Weihrauch received his PhD from the University of Bonn in 1973, and after a research associateship at Cornell University and a professorship position at RWTH Aachen he held a chair for Theoretical Computer Science at the University of Hagen from 1979 until his retirement in 2008. Together with his PhD students he developed the representation based approach to computable analysis, sometimes called *type-2 theory of effectivity*. Most notably among the early PhD students were Christoph Kreitz [68], Thomas Deil [30] and Norbert Müller [78]. A later generation of PhD students includes Peter Hertling [48], Xizhong Zheng [113], Matthias Schröder [95], the author of these notes, and others.[1]

---

*The author thanks Peter Hertling, Arno Pauly and Klaus Weihrauch for helpful comments on this article. This work has been supported by the *National Research Foundation of South Africa* (Grant Number 115269).

[1]A more complete Ph.D. genealogy of Weihrauch can be found in the preface of [21].

The main idea of this approach to computable analysis is to perform all computability considerations on Baire space, and to transfer theses concepts to other spaces by representing them with Baire space. On the one hand, on Baire space $\mathbb{N}^\mathbb{N}$ concepts such as continuity and computability are well understood, for instance with the help of Turing machines that operate on natural number sequences. On the other hand, natural number sequences can naturally be used to represent other objects such as real numbers, closed subsets or continuous functions on real numbers. Analogous statements hold for Cantor space $2^\mathbb{N}$, which additionally has natural concepts of time and space complexity.

Implicitly, such representations were used ever since Turing [102, 101] introduced his machines to operate on real numbers, and they are also implicit in constructive analysis [4], reverse mathematics [98], descriptive set theory [76, 63] and in set-theoretical constructions in classical mathematics too. However, the crucial idea of Weihrauch and his collaborators was to make such representations objects of mathematical investigations themselves by considering them as partial surjective maps.

**Definition 1.1** (Representation). A *representation* of a set $X$ is a partial surjective map $\delta :\subseteq \mathbb{N}^\mathbb{N} \to X$.[2]

This idea already had a tradition in mathematical logic since it was also present in Hauck's work [43, 44]. Kreitz and Weihrauch [70, 107, 71, 112] developed the theory of representations following the lines of the theory of numberings that was proposed in the 1970s by Ershov [35]. An early manifesto by Kreitz and Weihrauch can be found in [69] and more complete presentations in [108, 111].

It turned out that when dealing with infinite objects such as real numbers the proper choice of a representation is more crucial than for discrete objects. When one represents rational numbers $\mathbb{Q}$ one more or less automatically arrives at a suitable representation from the perspective of computability, and one has to work hard to find a representation that is not computably equivalent to a natural one, for instance by artificially encoding the halting problem into the representation.

In the case of infinite objects, such as real numbers $\mathbb{R}$, there are many natural representations that lead to mutually inequivalent structures. Figure 1 displays a portion of the lattice of real number representations that were studied already by Deil [30]. We omit the formal definitions, but we introduce some symbolic names in the diagram for later reference. The representations of the same shade of grey induce identical notions of computable real numbers: the Cauchy representation and all representations

---

[2]In this paper, the notation $f :\subseteq X \to Y$ abbreviates "$f$ is a function with $\mathrm{dom}(f) \subseteq X$ and $\mathrm{ran}(f) \subseteq Y$". Similarly, $f :\subseteq X \rightrightarrows Y$ abbreviates "$f$ is a multi-valued function with $\mathrm{dom}(f) \subseteq X$ and $\mathrm{ran}(f) \subseteq Y$", i.e., $f \subseteq X \times Y$.

```
                     Naive Cauchy representation
                              ρ'
                            ↗ ↖
 Enumeration         ρ<          ρ>         Enumeration
 of left cuts          ↖       ↗            of right cuts
                            ρ                Cauchy representation
                            ↑
                           ρ₁₀               Decimal representation
                          ↗ ↖
 Characteristic     ρcf<        ρcf>         Characteristic
 functions of left cuts  ↖    ↗              functions of right cuts
                           ρcf
                   Continued fraction representation
```

FIGURE 1. Lattice of real number representations.

below it in the diagram induce the ordinary notion of a computable real number; the representations via enumerations of left and right cuts induce the left- and right-computable real numbers, respectively; the naïve Cauchy representation induces the limit computable reals.

The preorder used in the diagram is that of reducibility of representations, which can be seen as a generalization of the concept of reducibility for numberings or of many-one reducibility to type-2 objects. Any arrow in the diagram indicates a reduction in the direction of the arrow, and all missing arrows indicate that no reduction is possible (except for those that follow by transitivity and reflexivity).

**Definition 1.2** (Reducibility). Let $f, g :\subseteq \mathbb{N}^{\mathbb{N}} \to X$ be partial functions. Then we say that *f is reducible to g*, in symbols $f \leq g$, if there exists a computable $F :\subseteq \mathbb{N}^{\mathbb{N}} \to \mathbb{N}^{\mathbb{N}}$ such that $f(p) = g \circ F(p)$ for all $p \in \text{dom}(f)$.

By $\equiv$ we denote the equivalence induced by $\leq$. One can also define a purely topological analogue of this reducibility by requiring that $F$ is continuous, and it turns out that the relations in the diagram in Figure 1 are not affected by this modification. Hence, one can say that the uniform distinctions between these representations are already of topological nature.

In particular, the continued fraction representation carries more continuously accessible information about real numbers than the decimal representation, and in turn the decimal representation carries more information

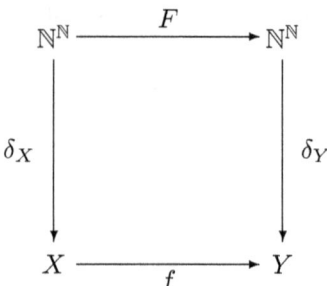

FIGURE 2. Continuity with respect to representations.

than the Cauchy representation. Having more informative representations on the input side is helpful, but it can be a burden on the output side. For instance addition is neither computable with respect to the continued fraction representation [66] nor with respect to the decimal representation [101].

These observations naturally lead to the question how one can identify a suitable representation among all the many representations of infinite objects that one can consider? The answer that Kreitz and Weihrauch gave is that a good representation has to be topologically natural, and the corresponding concept is called *admissibility*. Their main theorem on this topic is the following [69, 70, 108].

**Theorem 1.3** (Kreitz-Weihrauch 1984). *If $X$ and $Y$ are admissibly represented $T_0$–spaces with countable bases, then $f :\subseteq X \to Y$ is continuous if and only if it is continuous with respect to the underlying representations.*

Some further explanations are required here. For one, continuity with respect to the underlying representations means that the diagram in Figure 2 commutes.

That is, if $(X, \delta_X)$ and $(Y, \delta_Y)$ are represented spaces, then a function $f :\subseteq X \to Y$ is called *continuous with respect to the underlying representations*, if there is is a continuous $F :\subseteq \mathbb{N}^\mathbb{N} \to \mathbb{N}^\mathbb{N}$ such that $\delta_Y F(p) = f\delta_X(p)$ for all $p \in \text{dom}(f\delta_X)$. Other notions such as computability, Borel measurability, etc., can be transferred analogously to represented spaces, and Theorem 1.3 tells us that as long we use admissible representations, then at least topological continuity is the same as continuity with respect to the representations.

Theorem 1.3 was later generalized by Schröder [95] to a larger category of topological spaces, and one needs to replace topological continuity with sequential continuity in this more general context. Schröder also introduced a more general definition of admissibility that we are going to present here.

**Definition 1.4** (Admissibility). A representation $\delta :\subseteq \mathbb{N}^\mathbb{N} \to X$ of a topological space $X$ is called *admissible* if it is continuous and maximal among all continuous representations of $X$ with respect to the topological version of the reducibility $\leq$.

Among the real number representations it is the equivalence class of the Cauchy representation that yields admissible representations with respect to the Euclidean topology. In general, admissibility yields a handy criterion to judge whether a representation is suitable from a topological perspective. Many hyper and function space representations have been analyzed, such as representations for the space $\mathcal{C}(X)$ of continuous functions $f : X \to \mathbb{R}$, which we are not going to introduce in detail here [22, 111]. There are also other schools of computable analysis that approach the subject from a slightly different angle, such as the Pour-El and Richards school [92]. Studies of computational complexity in analysis were initiated by Norbert Müller [77, 79, 80], Ker-I Ko [67], and more recently by Akitoshi Kawamura and Stephen Cook [61, 62]. A more comprehensive discussion of the historical developments in computable analysis that arose out of Turing's work is presented in [2]. The tutorial [22] contains a concise introduction to computable analysis, and the handbook [20] contains surveys on many fascinating aspects of the more recent developments in computable analysis in general. From 1995 until today the computable analysis group founded in Hagen runs a conference with the title *Computability and Complexity in Analysis* (CCA). Some participants of the first meeting CCA 1995 in Hagen are shown in Figure 3.

## 2 Weihrauch reducibility

When studies in computable analysis progressed, the need arose to include the study of multi-valued maps. For instance, when one discusses the problem of finding zeros of a continuous function $f : X \to \mathbb{R}$, then it is not sufficient to consider single-valued functions of type $Z :\subseteq \mathcal{C}(X) \to \mathbb{R}$ in oder to describe zero finding, because under certain assumptions it might be possible to compute a zero of a continuous function $f : X \to \mathbb{R}$ only in a non-extensional way, i.e., such that the zero depends on the description of $f$ and not just on $f$ itself [111]. This phenomenon is best captured by describing zero finding as a multi-valued partial function of type $Z :\subseteq \mathcal{C}(X) \rightrightarrows \mathbb{R}$ (cf. Footnote 2). In general, many mathematical problems can be construed as such multi-valued maps. Hence we use the following general definition.

**Definition 2.1** (Problem). A *problem* $f :\subseteq X \rightrightarrows Y$ is a partial multi-valued map on represented spaces $X, Y$.

If we have two problems $f, g :\subseteq X \rightrightarrows Y$ of the same type, then we say that $f$ *refines* $g$, in symbols $f \sqsubseteq g$, if $\mathrm{dom}(g) \subseteq \mathrm{dom}(f)$ and $f(x) \subseteq g(x)$

FIGURE 3. A group picture with some participants of CCA 1995. Names from left to right: Klaus Weihrauch, Norbert Müller, Ludwig Staiger, Kostas Skandalis, Markus Bläser, Ker-I Ko, Peter Hertling, Jens Blanck, Matthias Schröder, Arthur Chou, Klaus Meer, Martin Hötzel Escardo, Pietro Di Gianantonio, Uwe Mylatz, Holger Schulz, Rudolf Freund, Janos Blazi, Achim Kallweit. Picture: Vasco Brattka.

for all $x \in \mathrm{dom}(g)$. Problems can naturally be combined in several ways. For instance, the *composition* $g \circ f$ of two problems $f :\subseteq X \rightrightarrows Y$ and $g :\subseteq Y \rightrightarrows Z$ is defined by $\mathrm{dom}(g \circ f) := \{x \in \mathrm{dom}(f) : f(x) \subseteq \mathrm{dom}(g)\}$ and

$$(g \circ f)(x) := \{z \in Z : (\exists y \in f(x))\ z \in g(y)\}$$

for all $x \in \mathrm{dom}(g \circ f)$. The particular choice of the domain is important, as this ensures that composition preserves computability and continuity in the appropriate way. Another operation on the problems $f :\subseteq X \rightrightarrows Y$ and $g :\subseteq W \rightrightarrows Z$ is the *product* $f \times g$ that is defined by $\mathrm{dom}(f \times g) := \mathrm{dom}(f) \times \mathrm{dom}(g)$ and

$$(f \times g)(x, w) := f(x) \times g(w)$$

for all $(x, w) \in \mathrm{dom}(f \times g)$.

As soon as problems are captured as multi-valued maps, there arises the need for a tool to compare the computational power of problems beyond refinement. In some sense, the concept of reducibility given in Definition 1.2 already gives us a way to compare (single-valued) problems of a certain type. However, only the input is subject to a pre-processing step here. Weihrauch's ideas of extending the notion of many-one reducibility such that also a post-processing of the output is considered, were laid out in two unpublished technical reports [109, 110]:

Klaus Weihrauch, *The degrees of discontinuity of some translators between representations of the real numbers*, Technical Report TR-92-050, International Computer Science Institute, Berkeley, July 1992.

Klaus Weihrauch, *The TTE-interpretation of three hierarchies of omniscience principles*, Informatik Berichte 130, FernUniversität Hagen, September 1992.

In his original definition, Weihrauch did not present the definition of his reducibilities in the way it is seen most often now. In order to make the definition as clear and simple as possible, we present a slightly more general definition in almost categorical terms that can be interpreted in different ways. By $\mathrm{id} : \mathbb{N}^{\mathbb{N}} \to \mathbb{N}^{\mathbb{N}}$ we denote the identity on Baire space.

**Definition 2.2** (Weihrauch reducibility). Let $C$ be a class of problems, and let $f, g$ be two problems. We introduce the following terminology:

1. $f$ is *strongly Weihrauch reducible* to $g$ (*with respect to $C$*), in symbols $f \leq_{\mathrm{sW}} g$, if there are $H, K \in C$ such that $H \circ g \circ K \sqsubseteq f$.

2. $f$ is *Weihrauch reducible* to $g$ (*with respect to $C$*), in symbols $f \leq_{\mathrm{W}} g$, if there are $H, K \in C$ such that $H \circ (\mathrm{id} \times g) \circ K \sqsubseteq f$.

The set $C$ has to be fixed in order to make the symbolic notation meaningful.

In the usual definition, we use for $C$ the class of all computable (or continuous) problems.[3] And this is what usually is called (the topological version of) Weihrauch reducibility. If $C$ is sufficiently nice, i.e., closed under composition and under product with id, then the above definitions actually yield preorders, no matter what $C$ is. And in fact, versions of Weihrauch reducibility have also been considered for the classes $C$ of (suitably defined) polynomial-time computable problems [62], arithmetic problems [75], or even hyperarithmetic piecewise computable problems [42]. In a more categorical setting concepts similar to Weihrauch reducibility were also studied by Hirsch [55], and in complexity theory similar concepts are known as (polynomial-time) many-one reducibility for functions on discrete spaces [62].

The intuitive idea of (strong) Weihrauch reducibility is illustrated in the diagrams in Figure 4. The idea is that $f$ being *strongly Weihrauch reducible* to $g$ means that $g$ composed with some suitable pre-processor $K$ and some suitable post-processor $H$ refines $f$. Suitability includes that the pre- and post-processors have to be from the set $C$. In the case of the ordinary (non-strong) reduction the post-processor additionally has access to the original input in form of some information determined by the pre-processor.

---

[3]That this actually yields the ordinary definition of Weihrauch reducibility follows from [25, Lemma 2.5].

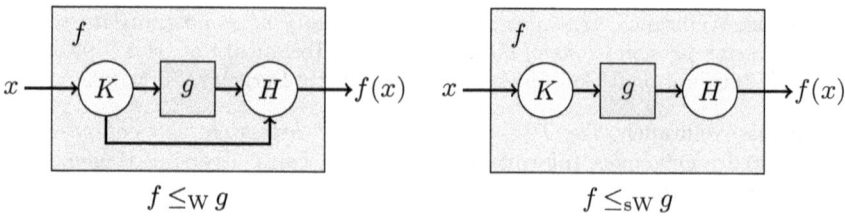

FIGURE 4. Weihrauch reducibility and strong Weihrauch reducibility.

In the original definition in [109, 110], Weihrauch firstly only considered single-valued problems $f, g$ on Cantor space, and for $C$ he used the set of continuous single-valued functions on Cantor space. He denoted the reducibilities $\leq_{\mathrm{sW}}$ and $\leq_{\mathrm{W}}$ by $\leq_2$ and $\leq_1$, respectively (and he used $\leq_0$ for a many-one like reducibility as in Definition 1.2). In a second step in [109, 110] he extended the definition to sets of such problems $f, g$. But this is just another technical way of dealing with multi-valuedness and essentially yields an approach that is equivalent to the modern one.[4] Hence, if not mentioned otherwise, we will assume from now on that $C$ is the class of all computable problems, and we will express everything in modern terminology. As usual, we denote the equivalences derived from strong and ordinary Weihrauch reducibility by $\equiv_{\mathrm{sW}}$ and $\equiv_{\mathrm{W}}$, respectively, and the strict versions of the reducibilities by $<_{\mathrm{sW}}$ and $<_{\mathrm{W}}$, respectively.

Only relatively late, it was discovered that the order structures induced by strong and ordinary Weihrauch reducibility are lattices. For $\leq_{\mathrm{W}}$ this result is due to the author and Gherardi [13] (for the lower semi-lattice) and Pauly [90] (for the upper semi-lattice). For $\leq_{\mathrm{sW}}$ the result is due to Dzhafarov [32].

**Proposition 2.3** (Weihrauch lattice). *The order structures induced by $\leq_{\mathrm{W}}$ and $\leq_{\mathrm{sW}}$ are lattices. In the case of $\leq_{\mathrm{W}}$ the lattice is distributive, in the case of $\leq_{\mathrm{sW}}$ it is not.*

More precise definitions along these lines and further results can be found in [17].

## 3  Weihrauch complexity

One goal of Weihrauch complexity is to classify the computational content of mathematical theorems of the logical form

$$(\forall x \in X)(x \in D \implies (\exists y \in Y)\, P(x, y))$$

---

[4]See the discussions in [25, Section 2.1] and [31, Appendix A].

| Weihrauch complexity | reverse mathematics |
|---|---|
| $C_1$ | $RCA_0$ without $\Sigma_1^0$–induction |
| $C_\mathbb{N}$ | $\Sigma_1^0$–induction |
| $C_{2^\mathbb{N}}$ | $WKL_0$ without $\Sigma_1^0$–induction |
| $C_\mathbb{R}$ | $WKL_0$ with $\Sigma_1^0$–induction |
| $C_{\mathbb{N}^\mathbb{N}}$ | $ATR_0$ |

FIGURE 5. Choice problems versus axiom systems.

in the Weihrauch lattice. If $X$ and $Y$ are represented spaces, then such a theorem directly translates into a Skolem-like problem

$$F :\subseteq X \rightrightarrows Y, x \mapsto \{y \in Y : P(x,y)\}$$

with $\mathrm{dom}(F) = D$. To locate the problem $F$ in the Weihrauch lattice amounts to classifying the uniform computational complexity of the corresponding theorem. One way to calibrate the complexity of some $F$ is to compare it to suitable *choice problems* for a suitable space $X$. By $C_X$ we denoted the *closed choice problem*

$$C_X :\subseteq \mathcal{A}_-(X) \rightrightarrows X, A \mapsto A$$

for a computable metric space $X$. The instances are non-empty closed sets $A \subseteq X$ represented by negative information (the space of all such closed sets is denoted by $\mathcal{A}_-(X)$) and the solution can be any point in $A$ (see [12, 17] for more precise definitions). Equivalently, we can see the instances as continuous functions $f : X \to \mathbb{R}$ with zeros and the solution can be any zero of the function $f$, i.e., if $\mathcal{C}(X)$ denotes the space of continuous function $f : X \to \mathbb{R}$, represented in a natural way, then

$$C_X :\subseteq \mathcal{C}(X) \rightrightarrows X, f \mapsto f^{-1}\{0\},$$

with $\mathrm{dom}(C_X) = \{f \in \mathcal{C}(X) : f^{-1}\{0\} \neq \varnothing\}$. This alternative description of choice shows that choice problems are basically about the solutions of equations of type

$$f(x) = 0$$

for continuous $f : X \to \mathbb{R}$. Typical spaces $X$ whose choice problems have been considered are Cantor space $X = 2^\mathbb{N}$, Baire space $X = \mathbb{N}^\mathbb{N}$, Euclidean space $X = \mathbb{R}$, the natural numbers $X = \mathbb{N}$, and finite spaces $n = \{0, ..., n-1\}$ for $n \in \mathbb{N}$.

Proving equivalences to choice problems roughly corresponds to a uniform version of a classification of the corresponding theorem in reverse mathematics [98]. Reverse mathematics is a proof-theoretic approach to classifying theorems according to which axioms are needed to prove the theorem in second-order arithmetic. Typical axiom systems are *recursive comprehension* $\mathsf{RCA}_0$, *arithmetic comprehension* $\mathsf{ACA}_0$, *arithmetic transitive recursion* $\mathsf{ATR}_0$. Details can be found in [98, 56]. The table given in Figure 5 indicates the correspondences between Weihrauch complexity and reverse mathematics.

This is just a very rough correspondence. For instance, $\mathsf{ATR}_0$ can be analyzed more closely in the vicinity of $\mathsf{C}_{\mathbb{N}^\mathbb{N}}$, and this is subject of several recent studies, for instance, by Marcone and Valenti [75], Goh, Pauly, and Valenti [41], Goh [40], Kihara, Marcone and Pauly [64].

The system $\mathsf{ACA}_0$ can be characterized using iterations of the *limit problem*

$$\lim {:}\subseteq \mathbb{N}^\mathbb{N} \to \mathbb{N}^\mathbb{N}, \langle p_0, p_1, p_2, ...\rangle \to \lim_{n\to\infty} p_n,$$

which is just the usual limit on Baire space (where for technical convenience, the input sequence is encoded by a standard tupling function $\langle\,\rangle$ in a single point in Baire space). Using these choice problems, one can obtain the following results. We are just mentioning some example for $\mathsf{C}_\mathbb{N}$, $\mathsf{C}_{2^\mathbb{N}}$ and lim. Many further results can be found in the survey [17]. The following result on problems equivalent to $\mathsf{C}_\mathbb{N}$ is due to the author and Gherardi [13].

**Theorem 3.1** (Choice on the natural numbers). The following are all Weihrauch equivalent to each other:

1. Choice on natural numbers $\mathsf{C}_\mathbb{N}$.

2. The Baire category theorem for computable complete metric spaces.

3. Banach's inverse mapping theorem for the Hilbert space $\ell_2$.

4. The open mapping theorem for $\ell_2$.

5. The closed graph theorem for $\ell_2$.

6. The uniform boundedness theorem on non-singleton computable Banach spaces.

In order to be more precise, one would have to explain which logical version of the theorem is translated into a problem here, and how the underlying data are represented. That these aspects can make a difference has been discussed for the Baire category theorem in detail [19]. We will not formalize these problems here and refer the interested reader to [13] for all details.

Next we mention a number of problems that are equivalent to $\mathsf{C}_{2^\mathbb{N}}$. The result on the Hahn-Banach theorem and the separation theorem is due to

Gherardi and Marcone [38], the result on the Brouwer fixed point theorem is due to the author, Le Roux, Joseph Miller, and Pauly [23]. The result on the Gale-Stewart theorem is due to Le Roux and Pauly [74]. All other results are easy to prove, the Heine-Borel theorem and the theorem of the maximum where briefly discussed in [8].

**Theorem 3.2** (Choice on Cantor space). *The following are all Weihrauch equivalent to each other:*

1. Choice on Cantor space $C_{2^\mathbb{N}}$.

2. Weak Kőnig's lemma WKL.

3. The Hahn-Banach theorem.

4. The separation theorem SEP on the separation of two disjoint enumerated sets in $\mathbb{N}$ by the characteristic function of another set (cf. below).

5. The Heine-Borel covering theorem.

6. The theorem of the maximum.

7. The Brouwer fixed point theorem for dimension $n \geq 2$.

8. The theorem of Gale-Stewart (on determinacy of games on Cantor space with closed winning sets).

Once again one would have to make these problems more precise. We give two examples. The *separation problem* is defined by

$$\mathsf{SEP} :\subseteq \mathbb{N}^\mathbb{N} \rightrightarrows 2^\mathbb{N}, \langle p, q \rangle \mapsto \{A \in 2^\mathbb{N} : \mathrm{ran}(p-1) \subseteq A \subseteq \mathbb{N} \setminus \mathrm{ran}(q-1)\}$$

with $\mathrm{dom}(\mathsf{SEP}) := \{\langle p,q\rangle : \mathrm{ran}(p-1) \cap \mathrm{ran}(q-1) = \varnothing\}$. Here $p-1 \in \mathbb{N}^* \cup \mathbb{N}^\mathbb{N}$ is the finite or infinite sequence of natural numbers that consists of the concatenation of

$$p(0) - 1, p(1) - 1, p(2) - 1, \ldots$$

with the understanding that $-1$ is the empty word. This technical construction is used to allow for enumerations of the empty set, and in order to keep 0 as dummy value in enumerations. We are going to see it later again.

Weak Kőnig's lemma is defined by

$$\mathsf{WKL} :\subseteq \mathrm{Tr} \rightrightarrows 2^\mathbb{N}, T \mapsto [T],$$

where Tr is the set of binary trees, $\mathrm{dom}(\mathsf{WKL})$ contains all infinite such trees, and $[T]$ is the set of infinite paths of such a tree $T$. As a simple example we mention at least one proof [13, Proposition 2.8 and Theorem 2.11]. If we represent the space $\mathcal{A}_-(2^\mathbb{N})$ of closed subsets of $2^\mathbb{N}$ by negative information, then the map $f : \mathrm{Tr} \to \mathcal{A}_-(2^\mathbb{N}), T \mapsto [T]$ is easily seen to be computable and it admits a multi-valued computable right inverse $g : \mathcal{A}_-(2^\mathbb{N}) \rightrightarrows \mathrm{Tr}$. This is all that is needed to prove $\mathsf{WKL} \equiv_\mathrm{W} \mathsf{C}_{2^\mathbb{N}}$.

We close this section with a number of problems that are equivalent to the limit operation. The results on the monotone convergence theorem and on the operator of differentiation $d$ (which is restricted to continuously differentiable functions) are due to von Stein [106] (see also Theorem 6.1). The Radon-Nikodym theorem was studied by Hoyrup, Rojas, and Weihrauch [58]. The other results are easy to show (see for instance [12, 8, 9]).

**Theorem 3.3** (The limit). The following are all Weihrauch equivalent to each other:

1. The limit map lim on Baire space (or Cantor space, or Euclidean space).

2. The Turing jump $\mathsf{J} : \mathbb{N}^\mathbb{N} \to \mathbb{N}^\mathbb{N}, p \mapsto p'$.

3. The monotone convergence theorem $\mathsf{MCT} :\subseteq \mathbb{R}^\mathbb{N} \to \mathbb{R}, (x_n)_{n\in\mathbb{N}} \mapsto \sup_{n\in\mathbb{N}} x_n$.

4. The operator of differentiation $d :\subseteq \mathcal{C}[0,1] \to \mathcal{C}[0,1], f \mapsto f'$.

5. The Fréchet-Riesz representation theorem for $\ell_2$.

6. The Radon-Nikodym theorem.

In between the degrees of $\mathsf{C}_{2^\mathbb{N}}$ and lim there is the degree of $\mathsf{C}_\mathbb{R}$ and the degree of the *lowness problem* $\mathsf{L} := \mathsf{J}^{-1} \circ \lim$. The following uniform version of the low basis theorem of Jockusch and Soare [60] was proved by the author, de Brecht and Pauly [12].

**Theorem 3.4** (Uniform low basis theorem). $\mathsf{C}_{2^\mathbb{N}} <_\mathrm{W} \mathsf{C}_\mathbb{R} <_\mathrm{W} \mathsf{L} <_\mathrm{W} \lim$.

This implies, in particular, $\mathsf{C}_\mathbb{N} <_\mathrm{W} \mathsf{L}$. There are many further studies, e.g., on Nash equilibria by Arno Pauly [89, 91] that can also be described with the help of choice, on probabilistic versions of choice, for instance, by the author and Pauly [24], the author, Gherardi and Hölzl [15], Bienvenu and Kuyper [3], on degrees defined by jumps of choice, e.g., by the author, Gherardi and Marcone [16], non-standard degrees obtained by Ramsey's theorem, by Dorais, Dzhafarov, Hirst, Mileti and Shafer [31], Hirschfeldt and Jockusch [56, 57], the author and Rakotoniaina [26], Dzhafarov, Goh, Hirschfeldt, Patey and Pauly [33], Cholak, Dzhafarov, Hirschfeldt and Patey [28], Marcone and Valenti [75], and Dzhafarov and Patey [34].

The classifications presented here are essentially in line with results from reverse mathematics [98]. There are several attempts to establish formal bridges between the proof-theoretic side of reverse mathematics to the more uniform computational side of Weihrauch complexity. These can be found, for instance, in work of Fujiwara [36], Uftring [104], and Kuyper [72].

## 4 Degrees of translators between real number representations

We are now going to discuss some of Weihrauch's early results from [109, 110]. We will try to put them into the context of more recent results with the aim to show that many important Weihrauch degrees were already anticipated in these initial reports. The main objective of [109] is the study of a kind of quotient structure of the diagram displayed in Figure 1. For any two representations $\delta_1, \delta_2$ of some set $X$ we can consider the *implication problem* $(\delta_1 \to \delta_2) :\subseteq \mathbb{N}^\mathbb{N} \rightrightarrows \mathbb{N}^\mathbb{N}$

$$(\delta_1 \to \delta_2)(p) := \{q \in \mathbb{N}^\mathbb{N} : \delta_2(q) = \delta_1(p)\}$$

with $\text{dom}(\delta_1 \to \delta_2) := \text{dom}(\delta_1)$, which measures the complexity of translating $\delta_1$ into $\delta_2$. In particular, $(\delta_1 \to \delta_2)$ is computable (in the sense that it has a computable refinement) if and only if $\delta_1 \leq \delta_2$. A benchmark that can be used to measure the complexity of other problems is the problem

$$\mathsf{EC} : \mathbb{N}^\mathbb{N} \to 2^\mathbb{N}, p \mapsto \text{ran}(p-1)$$

that translates an enumeration of a set into its characteristic function. Using this terminology one of the results obtained by Weihrauch is the following.

**Theorem 4.1** (Weihrauch [109, Theorem 4]). We obtain

$$\mathsf{EC} \equiv_\mathrm{W} (\rho' \to \rho) \equiv_\mathrm{W} (\rho' \to \rho_>) \equiv_\mathrm{W} (\rho_< \to \rho_>) \equiv_\mathrm{W} (\rho_< \to \rho).$$

The proof of this proposition builds on earlier results by von Stein [106]. While Theorem 4.1 describes quotients in the upper part of the diagram in Figure 1, the next result describes quotients that refer to the lower part of the diagram. In order to capture the complexity, one needs to modify the benchmark problem $\mathsf{EC}$ as follows. By $\mathsf{EC}_1 :\subseteq \mathbb{N}^\mathbb{N} \to 2^\mathbb{N}$ we denote the restriction of $\mathsf{EC}$ to such enumerations $p$ for which

$$|\mathbb{N} \setminus \text{ran}(p-1)| \leq 1$$

holds, i.e., the enumerated set $\text{ran}(p-1)$ is either $\mathbb{N}$ or $\mathbb{N} \setminus \{k\}$ for some $k \in \mathbb{N}$.

**Theorem 4.2** (Weihrauch [109, Theorem 22]). We obtain

$$\mathsf{EC}_1 \equiv_\mathrm{W} (\rho \to \rho_{\text{cf}}) \equiv_\mathrm{W} (\rho \to \rho_{\text{cf}>}) \equiv_\mathrm{W} (\rho_{\text{cf}<} \to \rho_{\text{cf}>}) \equiv_\mathrm{W} (\rho_{\text{cf}<} \to \rho_{\text{cf}}).$$

The proof of this result can be built on the observation that $\mathsf{EC}_1$ is equivalent to the rationality problem (see Proposition 7.1 below), and that relative to the rationality problem the continued fraction algorithm is computable. We mention one last result from [109] that is related to the separation problem. Similarly as with $\mathsf{EC}$, the separation problem $\mathsf{SEP}$ has a version $\mathsf{SEP}_1$ restricted to the separation of sets which are almost complementary in the sense that the pairs $\langle p,q\rangle \in \mathrm{dom}(\mathsf{SEP}_1)$ have to satisfy the additional condition

$$|\mathbb{N} \setminus (\mathrm{ran}(p-1) \cup \mathrm{ran}(q-1))| \leq 1.$$

Using this terminology, we can say something on the middle part of the diagram in Figure 1. Namely, the problem $\mathsf{SEP}_1$ characterizes exactly the complexity of translating the Cauchy representation into the decimal representation.

**Theorem 4.3** (Weihrauch [109, Theorem 13]). $\mathsf{SEP}_1 \equiv_\mathrm{W} (\rho \to \rho_{10})$.

There are further results along these lines included in [109] that we cannot all summarize here. The three Weihrauch degrees of $\mathsf{EC}$, $\mathsf{EC}_1$ and $\mathsf{SEP}_1$ have anticipated classes that have been studied much later. We mention some of these observations.

**Theorem 4.4.** We obtain:

1. $\mathsf{SEP} \equiv_\mathrm{W} \mathsf{WKL} \leq_\mathrm{W} \mathsf{EC} \equiv_\mathrm{W} \lim$,

2. $\mathsf{SEP}_1 \equiv_\mathrm{W} \mathsf{C}_{\#\leq 2} \leq_\mathrm{W} \mathsf{EC}_1 \equiv_\mathrm{W} \mathsf{SORT}$.

The equivalence $\mathsf{EC} \equiv_\mathrm{W} \lim$ follows from [7, Proposition] (see also [14, Lemma 5.3] and [12]). The equivalence $\mathsf{EC}_1 \equiv_\mathrm{W} \mathsf{SORT}$ is proved in Proposition 7.1 in the appendix, where also a definition of the problem $\mathsf{SORT}$ is given, which was originally introduced by Pauly and Neumann [85]. As mentioned before, the equivalence $\mathsf{SEP} \equiv_\mathrm{W} \mathsf{WKL}$ has been proved by Gherardi and Marcone [38, Theorem 6.7]. The reduction $\mathsf{WKL} \leq_\mathrm{W} \lim$ follows from the uniform version of the low basis theorem (Theorem 3.4) or can easily be proved directly.

Finally, $\mathsf{C}_{\#\leq 2}$ is the closed choice problem on Cantor space restricted to sets with one or two elements. This problem was introduced and studied extensively by Le Roux and Pauly [73]. The proof of $\mathsf{SEP}_1 \equiv_\mathrm{W} \mathsf{C}_{\#\leq 2}$ is sketched in Proposition 7.2 in the appendix. The fact $\mathsf{SEP}_1 \leq_\mathrm{W} \mathsf{EC}_1$ is stated by Weihrauch in [109, Section 6]. In the same section we also find the following statement (the first separation as conjecture).

**Proposition 4.5** (Weihrauch [109, Section 6]). $\mathsf{SEP}_1 <_\mathrm{W} \mathsf{EC}_1 <_\mathrm{W} \mathsf{EC}$.

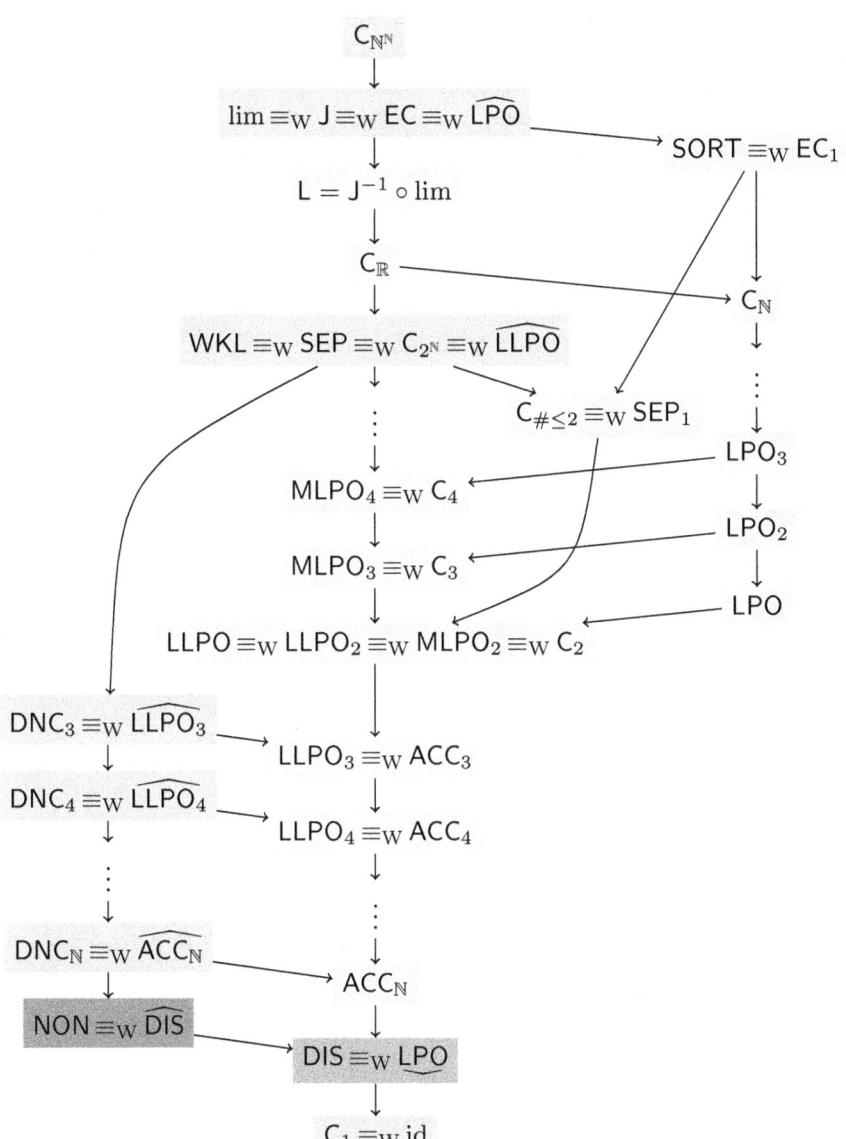

FIGURE 6. Some Weihrauch degrees.

Weihrauch attributes the separation in the second claim to Matthias Schröder. Indeed this follows from the fact that SORT always has a computable output, whereas EC can have non-computable outputs on computable inputs. The other separation $\mathsf{EC}_1 \not\leq_W \mathsf{SEP}_1$, mentioned as conjecture by Weihrauch, can be proved as follows: we have $\mathsf{LPO} \leq_W \mathsf{EC}_1$, but $\mathsf{LPO} \not\leq_W \mathsf{SEP}_1$, since otherwise $\lim \equiv_W \widehat{\mathsf{LPO}} \leq_W \mathsf{WKL}$ would follow, which is wrong by the low basis theorem (Theorem 3.4). The details and notations used in this proof will be discussed in the next section.

The diagram in Figure 6 shows a portion of the Weihrauch lattice that includes the problems discussed here together with problems mentioned in the next section. Any arrow indicates a Weihrauch reduction against the direction of the arrow (which is natural, as arrows correspond to logical implications in this way). The diagram should be complete up to transitivity and reflexivity.

## 5 Degrees of omniscience principles

In constructive mathematics principles of omniscience represent certain unsolvable problems of different degrees of complexity. Bishop introduced the *limited principle of omniscience* LPO and the *lesser limited principle of omniscience* LLPO [4]. Richman [94] generalized the latter mentioned principle to principles $\mathsf{LLPO}_n$. Weihrauch [110] translated these principles into problems in his lattice structure. We formulate them as multi-valued maps for all $n \geq 1$, where $\langle \rangle$ denotes some standard tupling function on Baire space:

$$\mathsf{LPO}_n : \mathbb{N}^{\mathbb{N}} \to \mathbb{N}, \ \langle p_1, ..., p_n \rangle \mapsto |\{i \in \mathbb{N} : p_i = \widehat{0}\}|,$$
$$\mathsf{LLPO}_n :\subseteq \mathbb{N}^{\mathbb{N}} \rightrightarrows \mathbb{N}, \ \langle p_1, ..., p_n \rangle \mapsto \{i \in \mathbb{N} : p_i = \widehat{0}\}$$
$$\text{with } \mathrm{dom}(\mathsf{LLPO}_n) := \{\langle p_1, ..., p_n \rangle : |\{i \in \mathbb{N} : p_i \neq \widehat{0}\}| \leq 1\},$$
$$\mathsf{MLPO}_n :\subseteq \mathbb{N}^{\mathbb{N}} \rightrightarrows \mathbb{N}, \ \langle p_1, ..., p_n \rangle \mapsto \{i \in \mathbb{N} : p_i = \widehat{0}\}$$
$$\text{with } \mathrm{dom}(\mathsf{MLPO}_n) := \{\langle p_1, ..., p_n \rangle : |\{i \in \mathbb{N} : p_i = \widehat{0}\}| \geq 1\}.$$

Using these general problems we define $\mathsf{LPO} := \mathsf{LPO}_1$ and $\mathsf{LLPO} := \mathsf{LLPO}_2 = \mathsf{MLPO}_2$. Among other results Weihrauch proved the following properties of these problems.

**Theorem 5.1** (Weihrauch [110, Theorems 4.2, 4.3, 5.4]). *For all $n, k \geq 1$ we obtain*

1. $\mathsf{LPO}_n <_W \mathsf{LPO}_{n+1}$,

2. $\mathsf{LLPO}_{n+2} <_W \mathsf{LLPO}_{n+1} <_W \mathsf{LPO}$,

3. $\mathsf{MLPO}_{n+1} <_W \mathsf{MLPO}_{n+2}$,

4. $\mathsf{MLPO}_{n+1} <_{\mathrm{W}} \mathsf{LPO}_n$,

5. $\mathsf{LPO}_k \not\leq_{\mathrm{W}} \mathsf{MLPO}_{n+1}$.

The diagram in Figure 6 illustrates these results. Later on, it was noticed that some of these omniscience proplems can also be seen as choice problems. By $\mathsf{C}_n$ we denote closed choice as defined in the previous section for the space $n = \{0, 1, ..., n-1\}$. By $\mathsf{ACC}_n$ we denote the *all-or-co-unique choice problem*, which is $\mathsf{C}_n$ restricted to sets $A \subseteq n$ with $|n \setminus A| \leq 1$. This definition is used analogously for $\mathsf{C}_\mathbb{N}$ and $\mathsf{ACC}_\mathbb{N}$ and the space $\mathbb{N} = \{0, 1, 2, ...\}$ of natural numbers. The following was observed in [12, Example 3.2] and [18, Fact 3.3].

**Proposition 5.2** (Choice and omniscience). We obtain $\mathsf{MLPO}_n \equiv_{\mathrm{W}} \mathsf{C}_n$ and $\mathsf{LLPO}_n \equiv_{\mathrm{W}} \mathsf{ACC}_n$ for all $n \geq 2$.

Some of the reductions and separations shown in Figure 6 have been proved elsewhere. We mention an example on choice and cardinality proved by Le Roux and Pauly [73, Figure 1].

**Proposition 5.3** (Le Roux and Pauly [73]). We obtain

1. $\mathsf{C}_2 <_{\mathrm{W}} \mathsf{C}_{\#\leq 2} \not\leq_{\mathrm{W}} \mathsf{C}_\mathbb{N}$,

2. $\mathsf{C}_3 \not\leq_{\mathrm{W}} \mathsf{C}_{\#\leq 2}$.

Other interesting results are related to the parallelization of omniscience problems. For any problem $f :\subseteq X \rightrightarrows Y$

$$\widehat{f} :\subseteq X^\mathbb{N} \rightrightarrows Y^\mathbb{N}, (x_n)_n \mapsto \underset{n \in \mathbb{N}}{\times} f(x_n)$$

is called the *parallelization* of $f$. This concept was introduced in [14], and it was proved that $f \mapsto \widehat{f}$ is a closure operator in the Weihrauch lattice. The following result characterizes the parallelizations of LPO and LLPO.

**Proposition 5.4** (B. & Gherardi [14, Theorem 8.2, Lemma 6.3]). We have: $\widehat{\mathsf{LLPO}} \equiv_{\mathrm{W}} \mathsf{WKL}$ and $\widehat{\mathsf{LPO}} \equiv_{\mathrm{W}} \mathsf{EC}$.

It was discovered by Higuchi and Kihara [54] that the parallelization of $\mathsf{LLPO}_n$ is related to the problems of *diagonally non-computable functions*

$\mathsf{DNC}_X : \mathbb{N}^\mathbb{N} \rightrightarrows X^\mathbb{N}, p \mapsto \{f \in X^\mathbb{N} : f \text{ is diagonally non-comp. relative to } p\}$,

defined for all $X \subseteq \mathbb{N}$. As usual, $f : \mathbb{N} \to X$ is called *diagonally non-computable relative to* $p$, if $(\forall i)\ \varphi_i^p(i) \neq f(i)$, where $\varphi_i^p$ is a Gödel numbering of the functions that are computable relative to $p \in \mathbb{N}^\mathbb{N}$. We now obtain the following parallelizations.

**Proposition 5.5** (Higuchi & Kihara [54, Proposition 81] and B., Hendtlass and Kreuzer [18, Theorem 5.2]). $\mathsf{DNC}_n \equiv_W \widehat{\mathsf{LLPO}_n}$ for all $n \geq 2$ and $\mathsf{DNC}_\mathbb{N} \equiv_W \widehat{\mathsf{ACC}_\mathbb{N}}$.

It is interesting to note that Jockusch [59] proved in 1989 a statement that is very similar to Weihrauch's result $\mathsf{LLPO}_{n+3} <_W \mathsf{LLPO}_{n+2}$ from Theorem 5.1 for the unparallelized case. Jockusch result was expressed in terms of Medvedev reducibility, but the separations imply separations of the corresponding Weihrauch degrees.

**Proposition 5.6** (Jockusch [59, Theorem 6]). We obtain for all $n \in \mathbb{N}$: $\mathsf{DNC}_\mathbb{N} <_W \mathsf{DNC}_{n+3} <_W \mathsf{DNC}_{n+2}$.

In fact, the results of Jockusch and Weihrauch can be jointly generalized to $\mathsf{ACC}_{n+2} \not\leq_W \mathsf{DNC}_{n+3}$ [18, Proposition 5.7].

For completeness, we have also added to Figure 6 the lowest known natural discontinuous Weihrauch degree, namely that of the *discontinuity problem* $\mathsf{DIS}$ [11]. This problem can be defined by *stashing* $\underline{\mathsf{LPO}}$ of $\mathsf{LPO}$ [10] (stashing is an operation dual to parallelization). We mention that the *non-computability problem* $\mathsf{NON}$, which was introduced and studied in [18], and maps every $p \in \mathbb{N}^\mathbb{N}$ to some $q \in \mathbb{N}^\mathbb{N}$ that is not computable from $p$, is the parallelization of $\mathsf{DIS}$ [10]. We omit the definitions here and point the interested reader to the references.

The main purpose of the diagram in Figure 6 is to illustrate that some of the problems discussed by Weihrauch in his initial two reports are related to important cornerstones among the Weihrauch degrees, some of which were much later rediscovered and studied under different names. All parallelizable and stashable degrees in Figure 6 are displayed in very dark grey, all only parallelizable degrees are displayed in mid grey, all only stashable degrees in dark grey, and all other degrees in light grey.

# 6 Some work by students in Hagen

The purpose of this section is to briefly discuss some follow-up work on Weihrauch's initial reports [109, 110]. In fact, Weihrauch supervised six M.Sc. and Ph.D. theses on topics related to his reducibility over a period of 18 years. The first of these even appeared before [109, 110] and are listed here in chronological order. Like [109, 110], most of this work remained unpublished for quite some time, and we give some references to later publications that overlap with the content of these theses.

*Diplomarbeit* of Torsten von Stein [106]. In this thesis, the concept of Weihrauch reducibility appears in writing for the first time and the author introduces the so-called C-hierarchy that measures the complexity of a problem $f$ by the number of applications of the problem $\mathsf{EC}$ that are required to solve $f$. He also studies quite a number of specific problems from analysis, most notably the differentiation problem (see Theorem 6.1 below). The results remain unpublished except for some pointers to this work elsewhere.

*Diplomarbeit* of Uwe Mylatz [81]. Uwe Mylatz continued the work of von Stein, and he studied also the second and third level of the C-hierarchy. Among other problems, he looked at higher derivatives, rationality of reals, monotonicity of sequences, convergence, density etc. This work remained unpublished.

*Diplomarbeit* of the present author [5]. In this thesis, it is essentially proved that the C-hierarchy is identical to the effective Borel hierarchy. A much extended version of this thesis was later published in [7], which also includes a proof that for admissible representations of computable Polish spaces Borel measurability on a certain finite level corresponds to the corresponding measurability of the realizers.

Ph.D. thesis of Peter Hertling [48]. The author introduces methods of combinatorial nature that allow to classify the (strong) topological Weihrauch degrees of certain problems with finite or countable discrete image, using forests of trees. These trees essentially capture the nature of the occurring discontinuities. Some of these results were already touched upon in the technical reports [47, 45, 46, 52] and the proceedings article [51]. A significant extension of some of the crucial results in [48] was finally published in [49]. This thesis has also influenced later work on this topic, for instance [97, 96, 50, 65].

Ph.D. thesis of Uwe Mylatz [82]. In his Ph.D. thesis, the author continues the study of the problems $\mathsf{LPO}_n, \mathsf{LLPO}_n, \mathsf{MLPO}_n$ and many variants thereof of combinatorial nature. Partially the methods developed in [48] are considered for separation results. The thesis remains unpublished.

M.Sc. thesis of Arno Pauly [88]. In this thesis, the author studies the classification of a number of concrete problems and the structure of the lattice induced by the continuous version of Weihrauch reducibility from a lattice-theoretic perspective. The publication [90] extends many of the structural results.

Presumably, the first time Weihrauch reducibility appeared in a published article was in [6], where the author generalized Pour-El and Richards first main theorem [92]. For instance the fact that the operator $d$ of differentiation is linear, has a closed graph and satisfies some minimal computability properties implies already the direction $\mathsf{EC} \leq_\mathrm{W} d$ in the following result.

**Theorem 6.1** (von Stein [106, Theorem 12]). *Let $d$ be the operator of differentiation $d :\subseteq \mathcal{C}[0,1] \to \mathcal{C}[0,1], f \mapsto f'$, restricted to continuously differentiable functions. Then $\mathsf{EC} \equiv_\mathrm{W} d$.*

## 7 Conclusion

The purpose of these notes is to show how the subject of Weihrauch complexity emerged in work of the Hagen school of computable analysis, in particular, from Weihrauch's original contributions to this topic and work produced by his students. Many more recent developments and prominent degrees were already anticipated in this early work.

Today Weihrauch complexity is a mature research topic that has found interest from researchers in computability theory, computable analysis, reverse mathematics, and proof theory. This is witnessed by two Dagstuhl seminars in 2015 and 2018 that were dedicated to Weihrauch complexity and a number of international Ph.D. and M.Sc. theses that are either mostly or partially dedicated to studying Weihrauch complexity. This includes the theses by Gherardi [37], Pauly [91], Higuchi [53], Carroy [27], Neumann [83, 84], Rakotoniaina [93], Borges [29], Patey [87], Sovine [99], Nobrega [86], Thies [100], Uftring [103], Goh [39], Anglés d'Auriac [1], and Valenti [105] (in chronological order from 2011 onwards, not mentioning the theses discussed in the previous section). A comprehensive up-to-date bibliography on Weihrauch complexity can be found online.

## Bibliography

[1] P.-E. Anglès d'Auriac. *Infinite Computations in Algorithmic Randomness and Reverse Mathematics*. Ph.D. thesis, Université Paris-Est, 2020.

[2] J. Avigad and V. Brattka. Computability and analysis: the legacy of Alan Turing. In R. Downey, editor, *Turing's Legacy: Developments from Turing's Ideas in Logic*, volume 42 of *Lecture Notes in Logic*, pages 1–47. Cambridge University Press, Cambridge, 2014.

[3] L. Bienvenu and R. Kuyper. Parallel and serial jumps of Weak Weak Kőnig's Lemma. In A. Day, M. Fellows, N. Greenberg, B. Khoussainov, A. Melnikov, and F. Rosamond, editors, *Computability and Complexity: Essays Dedicated to Rodney G. Downey on the Occasion of His 60th Birthday*, volume 10010 of *Lecture Notes in Computer Science*, pages 201–217. Springer, Cham, 2017.

[4] E. Bishop. *Foundations of Constructive Analysis*. McGraw-Hill, New York, 1967.

[5] V. Brattka. Grade der Nichtstetigkeit in der Analysis. Diplomarbeit, FernUniversität Hagen, 1993.

[6] V. Brattka. Computable invariance. *Theoretical Computer Science*, 210:3–20, 1999.

[7] V. Brattka. Effective Borel measurability and reducibility of functions. *Mathematical Logic Quarterly*, 51(1):19–44, 2005.

[8] V. Brattka. Computability and analysis, a historical approach. In A. Beckmann, L. Bienvenu, and N. Jonoska, editors, *Pursuit of the Universal. 12th Conference on Computability in Europe, CiE 2016, Paris, France, June 27–July 1, 2016. Proceedings*, volume 9709 of *Lecture Notes in Computer Science*, pages 45–57, Springer, Cham, 2016.

[9] V. Brattka. A Galois connection between Turing jumps and limits. *Logical Methods in Computer Science*, 14(3:13):1–37, Aug. 2018.

[10] V. Brattka. Stashing-parallelization pentagons. *Logical Methods in Computer Science*, 17(4):20:1–20:29, 2021.

[11] V. Brattka. The discontinuity problem. *Journal of Symbolic Logic*, accepted for publication.

[12] V. Brattka, M. de Brecht, and A. Pauly. Closed choice and a uniform low basis theorem. *Annals of Pure and Applied Logic*, 163:986–1008, 2012.

[13] V. Brattka and G. Gherardi. Effective choice and boundedness principles in computable analysis. *Bulletin of Symbolic Logic*, 17(1):73–117, 2011.

[14] V. Brattka and G. Gherardi. Weihrauch degrees, omniscience principles and weak computability. *Journal of Symbolic Logic*, 76(1):143–176, 2011.

[15] V. Brattka, G. Gherardi, and R. Hölzl. Probabilistic computability and choice. *Information and Computation*, 242:249–286, 2015.

[16] V. Brattka, G. Gherardi, and A. Marcone. The Bolzano-Weierstrass theorem is the jump of weak Kőnig's lemma. *Annals of Pure and Applied Logic*, 163:623–655, 2012.

[17] V. Brattka, G. Gherardi, and A. Pauly. Weihrauch complexity in computable analysis. In V. Brattka and P. Hertling, editors, *Handbook of Computability and Complexity in Analysis, Theory and Applications of Computability*, pages 367–417. Springer, Cham, 2021.

[18] V. Brattka, M. Hendtlass, and A. P. Kreuzer. On the uniform computational content of computability theory. *Theory of Computing Systems*, 61(4):1376–1426, 2017.

[19] V. Brattka, M. Hendtlass, and A. P. Kreuzer. On the uniform computational content of the Baire category theorem. *Notre Dame Journal of Formal Logic*, 59(4):605–636, 2018.

[20] V. Brattka and P. Hertling, editors. *Handbook of Computability and Complexity in Analysis, Theory and Applications of Computability*. Springer, Cham, 2021.

[21] V. Brattka, P. Hertling, K.-I. Ko, and N. Zhong, editors. *Computability and Complexity in Analysis. Selected Papers of the International Conference CCA 2003, held in Cincinnati, Ohio, August 28–30, 2003*. issue 4–5 of volume 50 of *Mathematical Logic Quarterly*, 2004.

[22] V. Brattka, P. Hertling, and K. Weihrauch. A tutorial on computable analysis. In S. B. Cooper, B. Löwe, and A. Sorbi, editors, *New Computational Paradigms: Changing Conceptions of What Is Computable*, pages 425–491. Springer, New York, 2008.

[23] V. Brattka, S. Le Roux, J. S. Miller, and A. Pauly. Connected choice and the Brouwer fixed point theorem. *Journal of Mathematical Logic*, 19(1):1–46, 2019.

[24] V. Brattka and A. Pauly. Computation with advice. In X. Zheng and N. Zhong, editors, *CCA 2010, Proceedings of the Seventh International Conference on Computability and Complexity in Analysis*, volume 24 of *Electronic Proceedings in Theoretical Computer Science*, pages 41–55, Open Publishing Association, 2010.

[25] V. Brattka and A. Pauly. On the algebraic structure of Weihrauch degrees. *Logical Methods in Computer Science*, 14(4:4):1–36, 2018.

[26] V. Brattka and T. Rakotoniaina. On the uniform computational content of Ramsey's theorem. *Journal of Symbolic Logic*, 82(4):1278–1316, 2017.

[27] R. Carroy. *Functions of the first Baire class*. Ph.D. thesis, Université de Lausanne & Université Paris Diderot—Paris 7, 2013.

[28] P. A. Cholak, D. D. Dzhafarov, D. R. Hirschfeldt, and L. Patey. Some results concerning the $\mathsf{SRT}^2_2$ vs. COH problem. *Computability*, 9(3–4):193–217, 2020.

[29] A. de Almeida Gabriel Vieira Borges. On the herbrandised interpretation for nonstandard arithmetic. M.Sc. thesis, Instituto Superior Técnico Lisbon, 2016.

[30] T. Deil. *Darstellungen und Berechenbarkeit reeller Zahlen*. Ph.D. thesis, FernUniversität Hagen, 1984.

[31] F. G. Dorais, D. D. Dzhafarov, J. L. Hirst, J. R. Mileti, and P. Shafer. On uniform relationships between combinatorial problems. *Transactions of the American Mathematical Society*, 368(2):1321–1359, 2016.

[32] D. D. Dzhafarov. Joins in the strong Weihrauch degrees. *Mathematical Research Letters*, 26(3):749–767, 2019.

[33] D. D. Dzhafarov, J. L. Goh, D. R. Hirschfeldt, L. Patey, and A. Pauly. Ramsey's theorem and products in the Weihrauch degrees. *Computability*, 9(2):85–110, 2020.

[34] D. D. Dzhafarov and L. Patey. COH, $SRT_2^2$, and multiple functionals. *Computability*, 10(2):111–121, 2021.

[35] J. L. Eršov. Theory of numberings. In E. R. Griffor, editor, *Handbook of Computability Theory*, volume 140 of *Studies in Logic and the Foundations of Mathematics*, pages 473–503. Elsevier, Amsterdam, 1999.

[36] M. Fujiwara. Weihrauch and constructive reducibility between existence statements. *Computability*, 10(1), 2021.

[37] G. Gherardi. *Some Results in Computable Analysis and Effective Borel Measurability*. Ph.D. thesis, University of Siena, 2006.

[38] G. Gherardi and A. Marcone. How incomputable is the separable Hahn-Banach theorem? *Notre Dame Journal of Formal Logic*, 50(4):393–425, 2009.

[39] J. L. Goh. *Measuring the Relative Complexity of Mathematical Constructions and Theorems*. Ph.D. thesis, Cornell University, August 2019.

[40] J. L. Goh. Embeddings between well-orderings: Computability-theoretic reductions. *Annals of Pure and Applied Logic*, 171(6):102789, 2020.

[41] J. L. Goh, A. Pauly, and M. Valenti. Finding descending sequences through ill-founded linear orders. *Journal of Symbolic Logic*, 86(2):817–854, 2021.

[42] N. Greenberg, R. Kuyper, and D. Turetsky. Cardinal invariants, non-lowness classes, and Weihrauch reducibility. *Computability*, 8(3–4):305–346, 2019.

[43] J. Hauck. Konstruktive Darstellungen reeller Zahlen und Folgen. *Zeitschrift für Mathematische Logik und Grundlagen der Mathematik*, 24:365–374, 1978.

[44] J. Hauck. Konstruktive Darstellungen in topologischen Räumen mit rekursiver Basis. *Zeitschrift für Mathematische Logik und Grundlagen der Mathematik*, 26:565–576, 1980.

[45] P. Hertling. Stetige Reduzierbarkeit auf $\Sigma^\omega$ von Funktionen mit zweielementigem Bild und von zweistetigen Funktionen mit diskretem Bild. Preprint, 1993 (Informatik Berichte 153, FernUniversität Hagen).

[46] P. Hertling. A topological complexity hierarchy of functions with finite range. In *Workshop on Continuous Algorithms and Complexity, Barcelona, October, 1993*. Preprint, 1993 (Centre de recerca matematica, Institut d'estudis catalans, Barcelona, Number 223).

[47] P. Hertling. Topologische Komplexitätsgrade von Funktionen mit endlichem Bild. Preprint, 1993 (Informatik Berichte 152, FernUniversität Hagen).

[48] P. Hertling. *Unstetigkeitsgrade von Funktionen in der effektiven Analysis*. Ph.D. thesis, FernUniversität Hagen, 1996.

[49] P. Hertling. Forests describing Wadge degrees and topological Weihrauch degrees of certain classes of functions and relations. *Computability*, 9(3-4):249–307, 2020.

[50] P. Hertling and V. Selivanov. Complexity issues for preorders on finite labeled forests. In V. Brattka, H. Diener, and D. Spreen, editors, *Logic, Computation, Hierarchies, Ontos Mathematical Logic*, pages 165–190. Walter de Gruyter, 2014.

[51] P. Hertling and K. Weihrauch. Levels of degeneracy and exact lower complexity bounds for geometric algorithms. In *Proceedings of the 6th Canadian Conference on Computational Geometry. Saskatoon, SK, Canada, August 1994*, pages 237–242. University of Saskatchewan, 1994.

[52] P. Hertling and K. Weihrauch. On the topological classification of degeneracies. Preprint, 1994 (Informatik Berichte 154, FernUniversität Hagen).

[53] K. Higuchi. *Degree Structures of Mass Problems and Choice Functions*. Ph.D. thesis, Tohoku University, 2012.

[54] K. Higuchi and T. Kihara. Inside the Muchnik degrees II: The degree structures induced by the arithmetical hierarchy of countably continuous functions. *Annals of Pure and Applied Logic*, 165(6):1201–1241, 2014.

[55] M. D. Hirsch. *Applications of topology to lower bound estimates in computer science*. Ph.D. thesis, University of California, Berkeley, 1990.

[56] D. R. Hirschfeldt. *Slicing the Truth: On the Computable and Reverse Mathematics of Combinatorial Principles*, volume 28 of *Lecture Notes Series, Institute for Mathematical Sciences, National University of Singapore*. World Scientific, Singapore, 2015.

[57] D. R. Hirschfeldt and C. G. Jockusch. On notions of computability-theoretic reduction between $\Pi_2^1$ principles. *Journal of Mathematical Logic*, 16(1):1650002, 59, 2016.

[58] M. Hoyrup, C. Rojas, and K. Weihrauch. Computability of the Radon-Nikodym derivative. *Computability*, 1(1):3–13, 2012.

[59] C. G. Jockusch, Jr. Degrees of functions with no fixed points. In J. E. Fenstad, I. T. Frolov, and R. Hilpinen, editors, *Logic, methodology and philosophy of science, VIII: proceedings of the Eighth International Congress of Logic, Methodology and Philosophy of Science, Moscow, 1987*, volume 126 of *Studies in Logic and the Foundations of Mathematics*, pages 191–201, North-Holland, Amsterdam, 1989.

[60] C. G. Jockusch, Jr. and R. I. Soare. $\Pi_1^0$ classes and degrees of theories. *Transactions of the American Mathematical Society*, 173:33–56, 1972.

[61] A. Kawamura. Lipschitz continuous ordinary differential equations are polynomial-space complete. *Computational Complexity*, 19(2):305–332, 2010.

[62] A. Kawamura and S. A. Cook. Complexity theory for operators in analysis. *ACM Transactions on Computation Theory*, 4(2):5:1–5:24, 2012.

[63] A. S. Kechris. *Classical Descriptive Set Theory*, volume 156 of *Graduate Texts in Mathematics*. Springer, Berlin, 1995.

[64] T. Kihara, A. Marcone, and A. Pauly. Searching for an analogue of $ATR_0$ in the Weihrauch lattice. *Journal of Symbolic Logic*, 85(3):1006–1043, 2020.

[65] T. Kihara and A. Montalbán. On the structure of the Wadge degrees of bqo-valued Borel functions. *Transactions of the American Mathematical Society*, 371(11):7885–7923, 2019.

[66] K.-I. Ko. On the continued fraction representation of computable real numbers. *Theoretical Computer Science*, 47:299–313, 1986. corr. in the same journal, 54 (1987), Pages 341–343.

[67] K.-I. Ko. *Complexity Theory of Real Functions*. Progress in Theoretical Computer Science. Birkhäuser, Boston, 1991.

[68] C. Kreitz. *Theorie der Darstellungen und ihre Anwendungen in der konstruktiven Analysis*. Ph.D. thesis, FernUniversität Hagen, 1984.

[69] C. Kreitz and K. Weihrauch. A unified approach to constructive and recursive analysis. In M. Richter, E. Börger, W. Oberschelp, B. Schinzel, and W. Thomas, editors, *Computation and Proof Theory. Proceedings of the Logic Colloquium, Aachen, July 18–23, 1983, Part II*, volume 1104 of *Lecture Notes in Mathematics*, pages 259–278, Springer, Berlin, 1984.

[70] C. Kreitz and K. Weihrauch. Theory of representations. *Theoretical Computer Science*, 38:35–53, 1985.

[71] C. Kreitz and K. Weihrauch. Compactness in constructive analysis revisited. *Annals of Pure and Applied Logic*, 36:29–38, 1987.

[72] R. Kuyper. On Weihrauch reducibility and intuitionistic reverse mathematics. *Journal of Symbolic Logic*, 82(4):1438–1458, 2017.

[73] S. Le Roux and A. Pauly. Finite choice, convex choice and finding roots. *Logical Methods in Computer Science*, 11(4):4:6, 31, 2015.

[74] S. Le Roux and A. Pauly. Weihrauch degrees of finding equilibria in sequential games (extended abstract). In A. Beckmann, V. Mitrana, and M. Soskova, editors, *Evolving Computability. 11th Conference on Computability in Europe, CiE 2015, Bucharest, Romania, June 29–July 3, 2015. Proceedings*, volume 9136 of *Lecture Notes in Computer Science*, pages 246–257, Springer, Cham, 2015.

[75] A. Marcone and M. Valenti. The open and clopen Ramsey theorems in the Weihrauch lattice. *Journal of Symbolic Logic*, 86(1):316–351, 2021.

[76] Y. N. Moschovakis. *Descriptive Set Theory*, volume 100 of *Studies in Logic and the Foundations of Mathematics*. North-Holland, Amsterdam, 1980.

[77] N. T. Müller. Subpolynomial complexity classes of real functions and real numbers. In L. Kott, editor, *Proceedings of the 13th International Colloquium on Automata, Languages, and Programming*, volume 226 of *Lecture Notes in Computer Science*, pages 284–293, Springer, Berlin, 1986.

[78] N. T. Müller. *Untersuchungen zur Komplexität reeller Funktionen*. Ph.D. thesis, FernUniversität Hagen, 1988.

[79] N. T. Müller. Polynomial time computation of Taylor series. In *Proceedings of the 22th JAIIO–Panel'93, Part 2*, pages 259–281, Buenos Aires, 1993.

[80] N. T. Müller and B. Moiske. Solving initial value problems in polynomial time. In *Proceedings of the 22th JAIIO—Panel'93, Part 2*, pages 283–293, Buenos Aires, 1993.

[81] U. Mylatz. Vergleich unstetiger Funktionen in der Analysis. Diplomarbeit, FernUniversität Hagen, 1992.

[82] U. Mylatz. *Vergleich unstetiger Funktionen: "Principle of Omniscience" und Vollständigkeit in der C-Hierarchie*. Ph.D. thesis, FernUniversität Hagen, 2006.

[83] E. Neumann. Computational problems in metric fixed point theory and their Weihrauch degrees. M.Sc. thesis, Technische Universität Darmstadt, 2014.

[84] E. Neumann. *Universal Envelopes of Discontinuous Functions*. Ph.D. thesis, Aston University, 2018.

[85] E. Neumann and A. Pauly. A topological view on algebraic computation models. *Journal of Complexity*, 44(Supplement C):1–22, 2018.

[86] H. Nobrega. *Games for functions - Baire classes, Weihrauch degrees, Transfinite Computations, and Ranks*. Ph.D. thesis, Universiteit van Amsterdam, 2018.

[87] L. Patey. *The reverse mathematics of Ramsey-type theorems*. Ph.D. thesis, Université Paris Diderot—Paris 7, 2016.

[88] A. Pauly. Methoden zum Vergleich der Unstetigkeit von Funktionen. M.Sc. thesis, FernUniversität Hagen, 2007.

[89] A. Pauly. How incomputable is finding Nash equilibria? *Journal of Universal Computer Science*, 16(18):2686–2710, 2010.

[90] A. Pauly. On the (semi)lattices induced by continuous reducibilities. *Mathematical Logic Quarterly*, 56(5):488–502, 2010.

[91] A. Pauly. *Computable Metamathematics and its Application to Game Theory*. Ph.D. thesis, University of Cambridge, 2011.

[92] M. B. Pour-El and J. I. Richards. *Computability in Analysis and Physics*. Perspectives in Mathematical Logic. Springer, Berlin, 1989.

[93] T. Rakotoniaina. *On the Computational Strength of Ramsey's Theorem*. Ph.D. thesis, University of Cape Town, 2015.

[94] F. Richman. Polynomials and linear transformations. *Linear Algebra and its Applications*, 131(1):131–137, 1990.

[95] M. Schröder. Extended admissibility. *Theoretical Computer Science*, 284(2):519–538, 2002.

[96] V. Selivanov. A fine hierarchy of $\omega$-regular $k$-partitions. In B. Löwe, D. Normann, I. N. Soskov, and A. A. Soskova, editors, *Models of Computation in Context. 7th Conference on Computability in Europe, CiE 2011, Sofia, Bulgaria, June 27–July 2, 2011. Proceedings*, volume 6735 of *Lecture Notes in Computer Science*, pages 260–269. Springer, Heidelberg, 2011.

[97] V. L. Selivanov. Hierarchies of $\Delta_2^0$–measurable $k$–partitions. *Mathematical Logic Quarterly*, 53(4–5):446–461, 2007.

[98] S. G. Simpson. *Subsystems of Second Order Arithmetic*. Perspectives in Mathematical Logic. Springer, Berlin, 1999.

[99] S. Sovine. Weihrauch reducibility and finite-dimensional subspaces. M.Sc. thesis, Marshall University, 2017.

[100] H. Thies. *Uniform computational complexity of ordinary differential equations with applications to dynamical systems and exact real arithmetic*. Ph.D. thesis, University of Tokyo, 2018.

[101] A. Turing. *Systems of Logic based on Ordinals*. Ph.D. thesis, Princeton University, 1938.

[102] A. M. Turing. On computable numbers, with an application to the Entscheidungsproblem. *Proceedings of the London Mathematical Society*, 42(2):230–265, 1937.

[103] P. Uftring. Proof-theoretic characterization of Weihrauch reducibility. M.Sc. thesis, Technische Universität Darmstadt, 2018.

[104] P. Uftring. The characterization of Weihrauch reducibility in systems containing E-PA$^\omega$ + QF-AC$^{0,0}$, *Journal of Symbolic Logic*, 86(1):224–261, 2021.

[105] M. Valenti. *A journey through computability, topology and analysis.* Ph.D. thesis, Universitá degli Studi di Udine, 2021.

[106] T. von Stein. Vergleich nicht konstruktiv lösbarer Probleme in der Analysis. Diplomarbeit, FernUniversität Hagen, 1989.

[107] K. Weihrauch. Type 2 recursion theory. *Theoretical Computer Science*, 38:17–33, 1985.

[108] K. Weihrauch. *Computability*, volume 9 of *EATCS Monographs on Theoretical Computer Science*. Springer, Berlin, 1987.

[109] K. Weihrauch. The degrees of discontinuity of some translators between representations of the real numbers. Preprint, 1992 (TR-92-050, International Computer Science Institute, Berkeley).

[110] K. Weihrauch. The TTE-interpretation of three hierarchies of omniscience principles. Preprint, 1992 (Informatik Berichte 130, FernUniversität Hagen).

[111] K. Weihrauch. *Computable Analysis*. Springer, Berlin, 2000.

[112] K. Weihrauch and C. Kreitz. Representations of the real numbers and of the open subsets of the set of real numbers. *Annals of Pure and Applied Logic*, 35:247–260, 1987.

[113] X. Zheng. *Weak Computability and Semi-Computability in Analysis.* Ph.D. thesis, FernUniversität Hagen, 1998.

## Appendix: Some Proofs

In this appendix we add some proofs that justify some claims made earlier. It seems that most of the corresponding equivalences stated here have not been noticed widely and do not appear in writing elsewhere.

We recall that the *sorting problem* SORT : $\{0,1\}^{\mathbb{N}} \to \{0,1\}^{\mathbb{N}}$ is defined by

$$\mathsf{SORT}(p) := \begin{cases} 0^n \widehat{1} & \text{if } p \text{ contains exactly } n \text{ zeros and} \\ \widehat{0} & \text{if } p \text{ contains infinitely many zeros.} \end{cases}$$

This problem was introduced by Pauly and Neumann [85]. Here $\widehat{n} = nnn... \in \mathbb{N}^{\mathbb{N}}$ denotes the constant sequence with value $n \in \mathbb{N}$.

We define the *rationality problem* RAT : $\mathbb{R} \rightrightarrows \{0,1\}^{\mathbb{N}}$ by

$$\mathsf{RAT}(x) := \begin{cases} \{0^n\widehat{1} : x = q_n\} & \text{if } x \in \mathbb{Q} \text{ and} \\ \{\widehat{0}\} & \text{if } x \in \mathbb{R} \setminus \mathbb{Q}, \end{cases}$$

where $(q_n)_{n \in \mathbb{N}}$ is the standard enumeration of the rational numbers $\mathbb{Q}$ given by $q_{\langle n,k,m \rangle} := \frac{n-k}{m+1}$, where $\langle \rangle$ denotes a standard Cantor tupling on $\mathbb{N}$. That is, RAT makes the property of being a rational number c.e. and on top of it, it determines the value of the rational number in the case of a rational input $x \in \mathbb{R}$. For irrational inputs $x$ no information is provided. Since the representations of the reals below and including $\rho$ in the diagram in Figure 1 are all computably equivalent, if restricted to irrational numbers, it is exactly the problem RAT that can help to translate the representations into each other. We prove the following statement about the equivalence class of RAT.

**Proposition 7.1.** $\mathsf{EC}_1 \equiv_{\mathrm{W}} \mathsf{RAT} \equiv_{\mathrm{W}} \mathsf{SORT}$.

*Proof.* We are going to prove $\mathsf{RAT} \leq_{\mathrm{sW}} \mathsf{EC}_1 \leq_{\mathrm{sW}} \mathsf{SORT} \leq_{\mathrm{W}} \mathsf{RAT}$.

We start with proving $\mathsf{RAT} \leq_{\mathrm{sW}} \mathsf{EC}_1$. Given an input $x \in \mathbb{R}$ for RAT, we start producing an enumeration $p \in \mathbb{N}^{\mathbb{N}}$ of $\mathbb{N} \setminus \{0\} = \mathrm{ran}(p-1)$ on the output side. Simultaneously, we try falsifying $x = q_0$. If it turns out that $x \neq q_0$, then we enumerate $0$ into our set and we switch to producing an enumeration of $\mathbb{N} \setminus \{m\}$ with some sufficiently large $m \in \mathbb{N}$ that has not yet been enumerated and such that $q_m = q_1$. Simultaneously, we try falsifying $x = q_m$. If we continue inductively in this way with some $m$ with $q_m = q_2$ in the next step, then for rational input $x \in \mathbb{Q}$ we produce some enumeration $p$ of a set $A = \mathbb{N} \setminus \{m\} = \mathrm{ran}(p-1)$ with $q_m = x$ and for irrational $x$ we produce some enumeration $p$ of $A = \mathbb{N} = \mathrm{ran}(p-1)$. Hence, given the characteristic function $\chi_A = \mathsf{EC}_1(p)$ we can compute a point in $\mathsf{RAT}(x)$ as follows: if $\chi_A = \widehat{1}$, then $\widehat{0} \in \mathsf{RAT}(x)$ and if $\chi_A = 1^m 0\widehat{1}$, then $0^m \widehat{1} \in \mathsf{RAT}(x)$.

We now prove $\mathsf{EC}_1 \leq_{\mathrm{sW}} \mathsf{SORT}$. Given $p \in \mathbb{N}^{\mathbb{N}}$ that enumerates a set $A = \mathrm{ran}(p-1)$ with $A = \mathbb{N}$ or $A = \mathbb{N} \setminus \{m\}$ for some $m \in \mathbb{N}$, we compute a sequence $q \in \{0,1\}^{\mathbb{N}}$ as follows. Whenever we have found a consecutive segment $\{0, 1, ...., k\} \subseteq A$ of length $k+1$ in the enumeration $p$, then we ensure that the output $q$ contains exactly $k+1$ zeros. As long as no other information is available on the input side, we append digits 1 to the output $q$. This algorithm ensures that $q$ contains infinitely many zeros (and hence $\mathsf{SORT}(q) = \widehat{0}$) if and only if $A = \mathbb{N}$ and $\mathsf{SORT}(q) = 0^m 111...$ if and only if $A = \mathbb{N} \setminus \{m\}$. Hence given $\mathsf{SORT}(q)$ we can compute the characteristic function $\chi_A$ of $A$ as follows: if $\mathsf{SORT}(q) = \widehat{0}$ then $\chi_A = \widehat{1}$ and if $\mathsf{SORT}(q) = 0^m \widehat{1}$ then $\chi_A = 1^m 0\widehat{1}$.

Now we prove $\mathsf{SORT} \leq_{\mathrm{W}} \mathsf{RAT}$. Given an input $p \in \{0,1\}^{\mathbb{N}}$ we start producing better and better approximations of $x_0 = \frac{1}{2^{\langle 0,0 \rangle}} \in \mathbb{R}$ on the output side. As soon as we find the first digit 0 in $p$, we switch to writing better and better approximations of some number $x_1 = \frac{2m+1}{2^{\langle 1,k \rangle}} \in \mathbb{R}$ on the output side with suitable $m, k \in \mathbb{N}$, such that $x_1$ is compatible with the previous approximations of $x_0$ and we start with producing an approximation of $x_1$ that is good enough such that no rational number with denominator smaller than $2^{\langle 1,k \rangle}$ can satisfy it (which is possible as $2m+1$ and $2^{\langle 1,k \rangle}$ are coprime). In general, if we find the $(n+1)$-th digit 0 in $p$, then we switch to producing approximations of some number $x_{n+1} = \frac{2m+1}{2^{\langle n+1,k \rangle}}$ for appropriate $m, k \in \mathbb{N}$, such that $x_{n+1}$ is compatible with the previous approximation of $x_n$. Again the first approximation of $x_{n+1}$ that we produce is good enough such that no rational number with denominator smaller than $2^{\langle n+1,k \rangle}$ can satisfy it. If there are infinitely many zeros in $p$, then the output converges to $x := \lim_{n \to \infty} x_n$, which must be irrational, since larger and larger denominators are excluded by the conditions above. If there is only a finite number $j$ of zeros in $p$, then the final output is $x = x_j$ and from $0^i\widehat{1} \in \mathsf{RAT}(x)$ we can compute $m, n, k \in \mathbb{N}$ such that $x = q_i = \frac{2m+1}{2^{\langle n,k \rangle}}$, where the numbers $m, n, k$ are unique since the numerator and denominator are coprime. Given $x$ and the original input $p$, we can now compute $\mathsf{SORT}(p)$ as follows: we write as many zeros to the output as we can find in $p$. Simultaneously, we try to find some $i$ with $0^i\widehat{1} \in \mathsf{RAT}(x)$. If we find such an $i$, then we determine $n$ as above and we extend the output to $0^n\widehat{1} = \mathsf{SORT}(p)$. If we never find such an $i$, then $p$ contains infinitely many zeros and the output is $\widehat{0} = \mathsf{SORT}(p)$.  Q.E.D.

The equivalence $\mathsf{RAT} \equiv_{\mathrm{W}} \mathsf{SORT}$ is similar to the statement of [85, Theorem 23].[5]

**Proposition 7.2** (Separation and choice). $\mathsf{C}_{\#\leq 2} \equiv_{\mathrm{sW}} \mathsf{SEP}_1$.

*Proof sketch.* Firstly, we note that

$$\mathrm{dom}(\mathsf{SEP}_1) = \{\langle p, q \rangle \in \mathbb{N}^{\mathbb{N}} : 1 \leq |\mathsf{SEP}\langle p, q \rangle| \leq 2\}$$

---

[5]However, [85, Theorem 23] does not seem to hold in full generality as stated, and the proof given here fills a gap and is an alternative proof for the case $X = \mathbb{R}$.

and negative information on the closed set $\mathsf{SEP}_1\langle p,q\rangle \subseteq 2^\mathbb{N}$ can be computed from $p,q$. This provides the reduction $\mathsf{SEP}_1 \leq_{\mathrm{sW}} \mathsf{C}_{\#\leq 2}$.

For the inverse reduction, we have given by negative information a closed set $A \subseteq 2^\mathbb{N}$ with one or two points. We can compute from the negative information an infinite binary tree $T$ with $[T] = A$. This tree has one or two infinite paths. We now follow the construction in the proof of [38, Theorem 6.7]. There a predicate $\varphi_T(s,i)$ is used, which indicates that for $s \in \{0,1\}^*$ and $i \in \{0,1\}$ there is a finite length $n \in \mathbb{N}$ such that the path $si$ can be extended in $T$ up to length $n$, but not the path $s(i-1)$. Without loss of generality, we assume that words $s \in \{0,1\}^*$ are identified with numbers $s \in \mathbb{N}$ that encode them with respect to some standard encoding. In the proof of [38, Theorem 6.7] it is shown how to compute $p,q \in \mathbb{N}^\mathbb{N}$ such that

$$\mathrm{ran}(p-1) = \{s+2 : \varphi_T(s,0)\} \cup \{0\},$$
$$\mathrm{ran}(q-1) = \{s+2 : \varphi_T(s,1)\} \cup \{1\}.$$

Since our tree $T$ has exactly one or two infinite paths, it follows that $\varphi_T(s,0) \vee \varphi_T(s,1)$ holds for all $s$, except possibly one. That is

$$|\mathbb{N} \setminus (\mathrm{ran}(p-1) \cup \mathrm{ran}(q-1))| \leq 1.$$

Hence, $\langle p,q\rangle \in \mathrm{dom}(\mathsf{SEP}_1)$, and the proof of the reduction $\mathsf{WKL} \leq_{\mathrm{sW}} \mathsf{SEP}$ in [38, Theorem 6.7] extends to a proof of the reduction $\mathsf{C}_{\#\leq 2} \leq_{\mathrm{sW}} \mathsf{SEP}_1$.

Q.E.D.

# Geschichte des Lehrstuhls für Logik und Grundlagenforschung an der Rheinischen Friedrich-Wilhelms-Universität Bonn

Elke Brendel & Rainer Stuhlmann-Laeisz

Institut für Philosophie, Rheinische Friedrich-Wilhelms-Universität Bonn, Am Hof 1, 53113 Bonn, Deutschland
E-Mail: {ebrendel,stuhlmann-laeisz}@uni-bonn.de

Im Juni 1959 beantragte die Philosophische Fakultät der Universität Bonn die Errichtung eines Ordinariats für Logik und Grundlagenforschung. Der Dekan der mathematisch-naturwissenschaftlichen Fakultät, Professor Roland Brinkmann, schrieb dazu am 3. Juli 1959:

> Die Logik und Grundlagenforschung gewinnen in der heutigen Wissenschaftssituation für die Mathematik und die Naturwissenschaften erhöhte grundsätzliche Bedeutung, die in Zukunft noch weiter zunehmen dürfte. Ein besonders lebhaftes Interesse an diesem Gebiet zeigt sich wohl bei der Reinen Mathematik (Ich verweise hier auf die s. Z. unter LORENZEN eingerichtete Abteilung für Grundlagenforschung der Mathematik beim Mathematischen Institut.) als auch bei der Angewandten Mathematik (logische Strukturen von Rechenanlagen, Kybernetik). Hier besteht sicher ein sehr reger Wunsch nach enger Zusammenarbeit mit einem solchen Lehrstuhl, der jedoch im Sinne der zusammenfassenden Idee der Universitas durchaus in der Philosophischen Fakultät beheimatet sein sollte.[1]

Auf den Lehrstuhl im neu errichteten Seminar für Logik und Grundlagenforschung sollte zunächst Paul Lorenzen berufen werden, der jedoch den Ruf im Jahr 1962 nicht annahm und stattdessen nach Erlangen ging, wie er in einem Schreiben an den Dekan vom 25. April 1962 erklärte:

> Spectabilität,
> für Ihre liebenswürdigen Bemühungen bin ich Ihnen zu großem Dank verpflichtet. Ich habe inzwischen von München ein endgültiges Angebot bekommen—und werde es in den nächsten Tagen annehmen. ... Die Gründe, aus denen ich Erlangen vorgezogen habe, sind persönlicher Art.[2]

Im Juni 1962 nimmt die Universität Bonn das Berufungsverfahren mit Gisbert Hasenjaeger auf, der den Ruf im selben Jahr annimmt. Zur Ge-

---

[1] Archiv der Universität Bonn, Aktenstelle MNF 88 287.
[2] Archiv der Universität Bonn, Anm. 5: Aktenstelle PF 77 Nr. 18.

ABBILDUNG 1. Gisbert Hasenjaeger im Jahre 1953. (Bild: Reinhold Remmert. Quelle: Bildarchiv des Mathematischen Forschungsinstituts Oberwolfach.)

schichte des durch die Berufung von Hasenjaeger begründeten Bonner Lehrstuhls für Logik und Grundlagenforschung schreibt Wolfram Hogrebe (Lehrstuhlinhaber für Theoretische Philosophie an der Universität von 1996 bis 2013):

> Hasenjaeger war Schüler von Heinrich Scholz in Münster und im Krieg nach 1942 in der Chiffrierabteilung des Oberkommandos der Wehrmacht beschäftigt. Ihm entging der Schwachpunkt der Chiffriermaschine Enigma, den Alan Turing für die Decodierung der militärischen Funksprüche der deutschen Wehrmacht nutzen konnte. Die Logik-Tradition in Bonn blieb seit Oskar Becker bis heute stabil, selbst wenn die Spannweite der philosophischen Fragestellungen natürlich variierte. Geforscht wurde insbesondere in Modallogik, epistemischer Logik, zu Frege und zu Problemen der Wissenschaftstheorie.[3]

Zu Hasenjaegers Schülern zählen viele bedeutende mathematischer Logiker, wie Ronald Jensen, Alexander Prestel, Dieter Rödding oder Peter Schroeder-Heister. Jensen promovierte 1964 bei Hasenjaeger in Bonn, wo er sich 1967 auch habilitierte. Er hatte u.a. von 1976 bis 1978 eine Professur in Bonn inne und war Professor für mathematische Logik von 1994 bis zu seiner Emeritierung im Jahr 2001 an der HU Berlin. Prestel habilitierte sich 1972 in Bonn. Nach einer Professur in Bonn 1973 wechselte er 1975 auf eine Professur an die Universität Konstanz. Rödding promovierte

---

[3]Wolfram Hogrebe: „Philosophie"; in: Thomas Becker, Philip Rosin (Hrsg.) Die Buchwissenschaften. Geschichte der Universität Bonn. Bd. 3. Bonn: Bonn University Press 2018. S. 683.

1961 bei Hasenjaeger und wurde als Nachfolger von Hans Hermes 1966 auf den Lehrstuhl für mathematische Logik und Grundlagenforschung an die Universität Münster berufen. Schroeder-Heister promovierte 1981 an der Universität Bonn bei Hasenjaeger (und Dag Prawitz, Stockholm). Er ist seit 1990 Professor für formale Logik und Sprachphilosophie an der Universität Tübingen und lehrt dort an der Philosophischen Fakultät sowie im Fachbereich Informatik.

Nach Hasenjaegers Emeritierung im Jahr 1984 wurde zum Sommersemester 1986 Rainer Stuhlmann-Laeisz auf den Bonner Logik-Lehrstuhl berufen. Stuhlmann-Laeisz, der von 1965 bis 1968 in Bonn studiert und bei Gisbert Hasenjaeger gehört hatte, war 1972 an der Universität Göttingen mit einer Dissertation über Kants Logik promoviert worden und hatte sich dort mit der Monographie *Das Sein-Sollen-Problem. Eine modallogische Studie* 1980 habilitiert. Mit diesen Qualifikationen war Stuhlmann-Laeisz als Kenner der philosophischen Logik ausgewiesen. Deren Verhältnis zur mathematischen Logik beschrieb er gern so:

> Gemeinsame Grundlage beider Disziplinen ist die an logischer Analyse der Sprache orientierte Normierung der Verwendungsweise logischer Konstanten. Während die mathematische Logik auf diesem Fundament ein überragendes Hochhaus errichtet, verbreitert die philosophische Logik das Fundament. Über die klassischen, extensionalen Konstanten hinaus werden auch solche sprachlichen Ausdrücke logisch normiert, die fundamental in philosophischen Theorien sind, exemplarisch die beiden intensionalen Modalitäten, notwendig zu sein bzw. gesollt zu sein.

Mit der Berufung von Stuhlmann-Laeisz war naturgemäß eine Schwerpunktverschiebung im Bereich der am Lehrstuhl erforschten Logik verbunden. Hinzu trat die Arbeit an Anwendungen der Logik auf klassische Fragen und Probleme der Philosophie und ihre Nutzung als Instrument zur Interpretation klassischer philosophischer Texte. Diese Ausrichtung schlug sich auch nieder in drei Habilitationen, nämlich in Ulrich Nortmanns Neuinterpretation der Modallogik des Aristoteles, in Albert Newens Monographie zur Philosophie des Geistes und in Thomas Müllers metaethischer Abhandlung über logische Strukturen in der normativen Theorie der Verpflichtung. Die drei Habilitierten sind inzwischen auf Lehrstühle für Philosophie berufen worden.

In die Dienstzeit von Stuhlmann-Laeisz vom 1. April 1986 bis zum 29. Februar 2008 fielen verschiedene organisatorische und örtliche Veränderungen des Lehrstuhls. Das Seminar für Logik und Grundlagenforschung war anfänglich untergebracht in zwei Etagen des Gebäudes in der Beringstrasse 1 (im Parterre die Bibliothek, die übrigen Räume im 1. Stock) in Bonn-Poppelsdorf, also ganz in der Nähe des Mathematischen Instituts der Universität und recht weit entfernt von deren Hauptgebäude, dem Schloss im Bonner Zentrum, dem Sitz der Philosophischen Seminare A und B. Als Institut war das Seminar—wie alle Universitätsinstitute—innerhalb der Philosophischen Fakultät eine hochschulrechtlich eigenständige Verwaltungseinheit mit einem Direktor.

Im Herbst 1989 erfolgte eine räumliche Veränderung: Das Seminar siedelte um in die Lennéstrasse 39. Dort stand neben zwei größeren Stockwerken auch das Souterrain zur Verfügung und die Entfernung zu den Philosophischen Seminaren im Hauptgebäude reduzierte sich auf sieben bis acht Gehminuten. Die inhaltliche Annäherung des Lehrstuhls an die Philosophie hatte damit auch räumlich ihren Niederschlag gefunden. Um die Jahrtausendwende erhoben sich Synergiebedenken gegen die Vielfalt der Universitätsinstitute in der Philosophischen Fakultät und insbesondere gegen die Existenz der kleineren unter diesen mit nur einer Professur. Es ergab sich eine Tendenz zur Neustrukturierung der Fakultät, die schließlich und erst nach Ablauf der Dienstzeit von Stuhlmann-Laeisz zur heutigen Gliederung in zehn Institute führte. Die Philosophie erfuhr diese Änderung in drei Schritten: Zuerst wurden die Seminare A und B zu einem Institut mit dem Namen *Philosophisches Seminar* verbunden. Später trat das Seminar für Logik und Grundlagenforschung diesem bei und gab damit seine Existenz als eigenständiges Universitätsinstitut auf. Damit erlosch dann auch sein Name. Die Widmung des Lehrstuhls als eines solchen für Logik und Grundlagenforschung und auch sein Name blieben erhalten. Die internen Bezeichnungen der vormaligen drei Seminare, A, B und C wurden ersetzt durch die Benennungen *Lehr- und Forschungsbereich 1* (bzw. *2* und *3*). Diese Gliederung hatte jedoch keine rechtlichen Implikationen, insbesondere waren die drei Bereiche keine Abteilungen des Instituts im verwaltungsrechtlichen Sinne. Die Umbenennung in Institut für Philosophie erfolgte später in einem dritten Schritt.

Nach der Emeritierung von Stuhlmann-Laeisz im Jahr 2008 wurde zum Wintersemester 2009/2010 der Lehrstuhl für Logik und Grundlagenforschung am Bonner Institut für Philosophie mit Elke Brendel neu besetzt. Brendel hat sich mit einer Arbeit über logisch-semantische Paradoxien an der Universität Frankfurt am Main promoviert und über Theorien des Wissens in der analytischen Erkenntnistheorie an der FU Berlin habilitiert. Bevor sie den Ruf an die Universität Bonn annahm, war sie Professorin für Philosophie mit dem Schwerpunkt Logik und Wissenschaftstheorie an der Universität Mainz. Ein zentrales Anliegen ihrer Forschungs- und Lehrtätigkeit am Bonner Logiklehrstuhl ist es, die Logik als traditionsreiche und eigenständige Teildisziplin der Philosophie sowie als Methode und Instrument der Analyse philosophischer Fragestellungen zu etablieren. Zu Brendels Arbeitsschwerpunkten am Bonner Logiklehrstuhl zählt insbesondere die Anwendung der Logik und formalen Semantik in der Metaphysik, Erkenntnis- und Sprachphilosophie. Sie untersucht beispielsweise Fragen nach der Natur, den Möglichkeiten und Grenzen des Wahrheitsbegriffs in formalen und natürlichen Sprachen. Sie arbeitet des Weiteren zur Modallogik und epistemischen Logik sowie zu Fragen nach der logisch-semantischen Struktur kontextsensitiver Ausdrücke. Ein zentrales Forschungsthema Brendels ist zudem die Philosophie der Logik sowie semantische Paradoxien in klassischen und nicht-klassischen Logiken. Brendel hat zu den genannten Arbeitsschwerpunkten zahlreiche Publikationen, internationale Kooperationen und Drittmittelprojekte realisiert. Sie leitet derzeit etwa ein Teilprojekt zum Thema „Abduktive Methoden in der Philosophie der Logik" im Rahmen einer DFG-Forschungsgruppe „Induktive Metaphysik" (Sprecher: Gerhard Schurz, Universität Düsseldorf). In diesem Projekt untersucht Brendel u.a. Fragen zum logischen Pluralismus, zu den Kriterien logischer Theoriewahl, zur Normativität der Logik sowie zum Verhältnis von Logik und den anderen Wissenschaften. Des Weiteren ist sie als eine der Herausgeberinnen und Herausgeber mit der Planung und Publikation des *Oxford Handbook of Philosophy of Logic* (Oxford University Press) befasst.

Die Logik ist eine wichtige und zentrale Säule des Lehrangebots am Bonner Institut für Philosophie. Im ersten Studienjahr des Bachelor-Studiums der Philosophie werden in einem obligatorischen Logik-Modul grundlegende Kenntnisse in der Syntax und Semantik der Aussagen- und Prädikatenlogik vermittelt sowie darauf aufbauend u.a. Kurse in Modallogik, nicht-klassischen Logiken, Argumentations- und Beweistheorie angeboten. Logikkenntnisse sensibilisieren Studierende für das Erkennen von Argumentstrukturen und helfen, logische Fehlschlüsse und argumentative Fallstricke zu vermeiden. Diese Logikkompetenzen werden dann im weiteren Verlauf des Bachelor-Studiums sowie im anschließenden Master-Studium durch weiterführende Seminare zu historischen und systematischen Aspekten der Logik vertieft. Viele Studierende wählen Themen aus dem Bereich der Logik für ihre Qualifikationsarbeiten. Neben Bachelor- und Masterarbeiten wurden und werden zudem zahlreiche Dissertationen, aber auch Habilitationen am Lehrstuhl für Logik und Grundlagenforschung betreut.

Auch während Brendels Zeit als Inhaberin des Bonner Logiklehrstuhls ergaben sich räumliche Veränderungen. Nachdem die Universität Bonn den Standort in der Lennéstraße 39 im Jahr 2013 aufgegeben hatte, musste der Lehrstuhl mitsamt der Bibliothek und allen Mitarbeiterinnen und Mitarbeitern in das Hauptgebäude der Universität umziehen. Im Jahr 2022 steht nun wiederum ein Umzug an. Da das kurfürstliche Schloss, der repräsentative Hauptsitz der Bonner Universität, vollständig saniert werden soll, müssen sämtliche Institute, die im Hauptgebäude beheimatet sind, auf zunächst unbestimmte Zeit auf andere Gebäude ausweichen. Der Bonner Logiklehrstuhl hat das Glück, zusammen mit anderen Philosophielehrstühlen sowie Lehrstühlen der Alten Geschichte in ein recht attraktives Gebäude in der Heinrich-von-Kleist-Straße in der Bonner Südstadt umgesiedelt zu werden. Auch in Zukunft wird die Logik nicht nur durch einen eigenen Lehrstuhl, sondern auch durch eine eigene Lehrstuhlbibliothek, die in der neuen Unterkunft erhalten bleiben soll, an der Universität Bonn eine wichtige Rolle spielen und weiterhin nationale wie internationale Strahlkraft ausüben.

# Logic and foundations of the exact sciences at the University of Konstanz: people & projects 1966–2021

Bernd Buldt

Department of Mathematical Sciences, Purdue University Fort Wayne, 2101 E. Coliseum Blvd., Fort Wayne IN 46805-1499, United States of America
E-mail: `buldtb@pfw.edu`

**Three remarks and a disclaimer**

Some remarks, I think, are in order before we begin.

(1) I decided to write in English at the off-chance that someone from outside the German-speaking countries may look at and scour this report for information. But to give it some fluency, I translated language that is tied to peculiarities of the German system of higher education quite freely.

German universities have traditionally been set up and operating in ways that are very different from other countries (e.g., the American tenure-track system). Faculty positions in Germany are internally known, and in their ramifications well understood, according to salary ranges: W1, W2, and W3 since 2002 and C1, C2, C3, and C4 before.[1]

Painting with a broad brush, we can say a W3 or C4 position is known as a chair (*Lehrstuhl*), its tenured incumbent is known as an *Ordinarius* or *Ordinaria*; the chair has its own administrative staff position(s) and one or more positions for postdoctoral academic staff who are assisting the chairholder. These assistant positions are untenured and fixed-term and used to be called *wissenschaftlicher Assistent* in the C-salary scales (until 2002). They are roughly the German equivalent of a tenure-track assistant professorship, except that they do not usually come with a path towards tenure. Chairholders wield a fair amount of power: the denomination of the chair gives its occupant (not the department) control over the curriculum in that field, they (not the department) hire their assistants, and they (not the department) accept and train their own graduate students. A W2 or C3 position, by contrast, is a tenured full professorship that comes with fewer of the perks. One innovation of the W-salary scales is that the holder of a W1 position is no longer an assistant to the chair, but a *Juniorprofessor*. These are non-tenured, fixed-term departmental professorial positions hired by a committee. In this article, we refer to researchers on both W1 and C1 positions, i.e., *Juniorprofessoren* and *Wissenschaftliche Assistenten* as

---

[1] We can ignore the former, post-war H-salary scales.

"assistant professors"; all other postdoctoral academic staff is called "postdoc".

While in the American system, faculty normally rise through the ranks of assistant, associate, and full professor at the same university by promotion, the German system does not know academic promotions; the default is that an external job offers are required to go from a W1 to a W2 position and then again from a W2 to a W3 position as chair.

In what follows I shall ignore these differences according to salary and endowment and simply refer to anyone on a W3, C4, W2, or C3 position as "professor" unless the context requires me to be more specific. The German system also knows tenured faculty on positions that are not of considered professorial rank (e.g., *wissenschaftlicher Mitarbeiter, Akademischer Rat, Akademischer Oberrat* etc.). I refer to those as "lecturer".

The qualification required to serve on a student's graduation committee is the *venia legendi*, which is normally obtained at the end of one's time as *wissenschaftlicher Assistent* by earning a second post-graduate degree, the *Habilitation*. Thus, while many faculty may participate in the education of masters and doctoral students, not all will have an official say in the matter.

Finally, we follow the German custom and identify a university with the city it is in. For example, when we say 'professor $X$ moved to Bochum,' we mean she accepted a position at the Ruhr-Universität Bochum; writing this out every time feels clunky. Note that we write "Konstanz" instead of "Constance;" the reason is that even those who hail from English-speaking countries almost always use the city's German name, not the English one.

(2) The report needs a modicum of structure. Thus, while I freely admit that drawing lines is highly arbitrary, I organized the report mainly by "era" (i.e., the tenure of faculty who hold a chair). This sounds a bit too pompous, and it probably is, but it does the job. In each of the sections, we structure the text by these chronological eras: for each era, we list in separate subsections first **People: their stories & their projects**, then **Teaching, research, students**, after that **Notable projects, conferences, & guests**, and finally **Selected publications**. In the list of publications, we only mention works that originated while their authors were in Konstanz and do not list items we already mentioned in the text. We prefer collections and books over journal articles and try to represent everyone involved.

(3) The overall emphasis of my report will be on logic and work on the foundations of the exact sciences insofar the latter employs formal methods or reflects on them. The limitation to the formal aspects is advised since work that was done in Konstanz on the foundations of the exact sciences in general is simply too rich and too multifaceted to be included *in toto*. People who made substantial and often sustained contributions to the philosophy of

the exact sciences, such as Michael Esfeld, Paul Hoyningen-Huene, Martin Carrier, Gertrude Hirsch Hadorn, David Hyder, Peter Janich, Klaus Mainzer, Johanna Seibt, Marcel Weber, Gereon Wolters, to name just a few, all held rank-and-file positions at the university but won't find much mention below. This still leaves the question what counts as logic or formal. The program of any recent *Colloquium Logicum* is certainly more inclusive than they were when the DVMLG met at Oberwolfach for their annual April meeting and held its general membership meeting there. For this report, I take guidance from recent meetings and include what I believe would be considered an appropriate topic for presentation; I admit that this is fuzzy and subjective.

(4) Finally, the disclaimer. I left Konstanz more than 15 years ago and moved too far away for a quick road trip to visit the university archives. So this report is primarily based on anecdotal evidence with little support from historical records or documents I could peruse. I sent inquiries to various witnesses of the time, and almost everyone was kind enough to answer in writing or to make time for an online meeting or a telephone interview. You know who you are: Thank you!!! But I could not close all gaps, nor is, as we all know, personal memory always a reliable source. I did my best to be a faithful chronicler. I also tried to balance the various sections, so that equitable weight is given to all players, and made serious efforts to find an appropriate mix of storytelling and adducing facts. The limitations I faced, however, are obvious; and I want them to be understood.

# 1 Introduction

The *Universität Konstanz*, like many other universities that sprang up around the same time in many countries around the globe, is the result of post-war, baby-boomer economic needs and began its operation in 1966. But it was conceived by the *Gründungsausschuss* (founding committee) as something special, namely, as a "reform university" (*Reformuniversität Konstanz*). Various moving pieces contributed to the idea of a reform university: an explicit focus on cutting-edge research; a commitment to Humboldt's idea of research-informed teaching where intimate seminars take the place of big lecture halls; an institution meant to be small (with just thirteen departments and without any of the engineering programs and no medical school) but elite ("Little Harvard on Lake Constance" was its nickname); the implementation of architectural as well as administrative structures aimed at fostering cooperation and rewarding creativity and innovation: an open architecture with a common hub of services (library, canteen) along with close physical proximity on a single shared campus was meant to facilitate contact across disciplinary borders while limited professional development monies should incentivize faculty to seek external grants.

We mention these points since they actually shaped how logic and philosophy of science were done. When it comes to logic, there was a fair amount of cooperation among linguists, mathematicians, and philosophers on campus but also with institutions nearby (notably, Freiburg, Stuttgart, Tübingen, and Zürich); cooperation with computer science seems more recent—computer science moved out of mathematics and became a department of its own in the academic year 1999/2000—and not frequent. There was a higher than average number of externally funded projects (e.g., the number of philosophy faculty funded by grant money easily dwarfed the number of regular faculty at any given point in time, while for linguistics it even meant more faculty than students, at least early on), and Konstanz saw international cooperations and guests whose numbers were out of proportion for its age or size. In regards to the foundations of the exact sciences it should be mentioned that philosophy of science has been part of the university's master plan before it even started its operation. The plan for a new, reform university had, in its early stage, included a proposal for what was called the *Interfakultät*. The idea was to have a certain pool of faculty members not form traditional departments in their disciplines of training (e.g., mathematics or philosophy) but to distribute them over the three main schools for which they would primarily offer service courses and, in case of philosophy, engage in a critical reflection of the disciplines housed in that school. The plan was not implemented, but its spirit lingered as we can observe on multiple occasions below.

## 2 Logic in linguistics

It may seem strange, but linguists did serious work in logic before everyone else did in Konstanz. The study of languages was institutionalized in a very peculiar form in Konstanz. It was split into two divisions: One was called 'literary studies' (*Literaturwissenschaften*), the other division was called 'linguistics' (*Sprachwissenschaften*). This was (except for Bielefeld) in stark contrast to any other university at the time, when the discipline was organized according to individual philologies (German, English, French, etc.) or language families (Romance languages, East-Asian languages, etc.). This design did not go back to the *Gründungsausschuss*, as rumor has it, but originated in the so-called 'Rhedaer Memorandum' which, born out of the spirit of the student unrest at the time, was an initiative to free philologies, and in particular German studies, from the shackles of the (Nazi) past. For Konstanz this meant that the study of language was much less siloed than at most of the other places.

**Era von Stechow (and Beyond): 1969–2008**

In respect to logic, the most consequential hire among first-generation faculty was Peter Hartmann (1923–1984; in Konstanz: 1969–1984). Hartmann came to Konstanz as an accomplished linguist but of a very traditional mold and with no track record of using formal methods; his dissertation was on Japanese grammar (1950) and his habilitation on nominal phrases in Sanskrit (1953), both with Alfred Schmitt in Münster whose successor he had become. At the time, however, linguistics at Münster was known to be among the more progressive departments, and so Hartmann brought along Klaus Brockhaus (1933–2011) as his assistant professor, who in turn had Arnim von Stechow (born 1941) in his tow as a post-doc. Beyond the brains, they came with a shiny new journal to publish their research in: *Linguistische Berichte*, founded by Hartmann and von Stechow in 1969. Brockhaus had majored in mathematics before he turned to formal linguistics; his doctoral work was on Carnap's *Aufbau*, with Hans Hermes in Münster (1963), and his habilitation was on automatic translations (1969). When Brockhaus left Konstanz already two years later, his former student von Stechow, who had done his doctoral work on finite-state machines with him (Münster, 1969), succeeded him as Hartmann's assistant and got a tenured position in 1972 which he held for the next 20 years (until he moved to nearby Tübingen in 1992).

**People: their stories & their projects.** When von Stechow and Brockhaus started their work, formal semantics did not yet exist as a discipline; worse, Brockhaus called von Stechow's ideas "interesting, but totally absurd." Their first approach was axiomatic (and hence more syntactic than model-theoretic) and oblivious of the work done elsewhere. Then, in 1971, Yehoshua Bar-Hillel spent his sabbatical in Konstanz. He applauded their early work in formal semantics but recommended Montague, or, in von Stechow's recollections: "he said he had studied the thing Brockhaus and I had written, and he said it was ingenious and it had gone almost so [sic!] far as Montague, and I should read that." Another seminal influence was a conference, organized by Edward Keenan in Cambridge (1973), with, among others, Hans Kamp, George Lackoff, David Lewis, and Barbara Partee in attendance and giving talks. Von Stechow switched gears, shifted more towards model-theoretic means, and assembled a research group that was destined to become a German, if not European, center for Montagovian formal semantics. A student of von Stechow, Peter (Eberhard) Pause—the first linguistics PhD in Konstanz (1972, on the logical complexity of transformation rules)—stayed on until his retirement (1974–2002), and in 1978 they were joined by Urs Egli (1941–2018; in Konstanz: 1978–2006) who, for lack of better opportunity, had written his dissertation on Stoic logic (Bern, 1967) but followed his early interests (Carnap, Chomsky, Montague) in his habilitation (Bern, 1973). There was hence a core group of three early-career faculty, supported by Hartmann, who worked on various aspects of semantics using mathematical and logical tools. And they were ably supported by gifted students. Those of their students who made a name for themselves in formal semantics or computational linguistics include—all earned their PhD in Konstanz except for Heim-Rainer Bäuerle (Stuttgart), Markus Egg (HU Berlin), Klaus von Heusinger (Cologne), Irene Heim (MIT), Angelika Kratzer (UMass), Uwe Mönnich (Tübingen), Wolfgang Sternefeldt (Tübingen), and (Thomas) Ede Zimmermann (Frankfurt). The latter's doctoral thesis on intensional type theory might very well be the shortest in recent memory, but it was von Stechow, not the candidate, who deemed 13 pages published in the *Journal of Symbolic Logic* (on an erroneous assumption undergirding Montague's work) sufficient.

It was a lucky conincidence that the DFG (*Deutsche Forschungsgemeinschaft*, the German Federal Research Agency) had identified linguistics as a focus area for special funding (*DFG Schwerpunktprogramm "Theoriebildung und Methodenentwicklung für die Linguistik"*, 1969–1974). In its context linguists at Konstanz established and received funding for their own special research area (SFB, *Sonderforschungsbereich*) which they called, somewhat pretentious it its simplicity: "linguistics." Internally it was known by its number: SFB 99. It ran from 1977 to 1985, and Hartmann served as its first speaker. But not only faculty at the rank of full professor, like Hartmann, or regular contributors such as Christoph Schwarze (Romance philology) were eligible to receive funding for individual projects within the SFB but also von Stechow and Egli, who at the time were not yet of professorial rank but on lecturer positions. (Egli's initial hire was actually a condition for receiving the SFB grant.) The SFB 99 continued under the title *Grammatik und sprachliche Prozesse* and was followed by the DFG-funded research group *Forschungsgruppe Theorie des Lexikons* (1986–1995), which led to another *Sonderforschungsbereich*: *Variation und Entwicklung im Lexikon* (SFB 471, 1997–2008), which also included individual projects by Egli (diachronic lexical semantics) and Pause (semantics of verbal phrases). Ample grant money allowed semanticists to organize international conferences which fostered international exchange and enabled them to stay at the cutting edge of their discipline; guests included Renate Bartsch, Max Cresswell, Hans Kamp, David Lewis, Terry Parsons, Barbara Partee, and Helmut Schnelle; Angelika Kratzer returned to Konstanz for conferences and Irene Heim as a Senior Fellow at the *Zukunftskolleg* (for which see § 5).

Their objects of study were all the more difficult topics that many instructors hide or gloss over when they teach first-order logic as the last word (quantifier behavior; (in)definite articles; nominal, verbal, and adverbial phrases; context sensitivity and anaphora; etc., you name it) and the tools they used were the usual suspects: formalized languages, automata theory, intensional and non-classical logics, type theory and $\lambda$-abstraction, as well as model-theoretic semantics. But they also conducted a quite detailed study of whether the $\iota$- or $\varepsilon$-operator of Russell and Hilbert, respectively, had promise for linguistics.

A project in the history of logic also came out of linguistics. Within the SFB 99, Egli had acquired funding for a project on the influence of Stoic dialectics (i. e., Stoic logic, philosophy of language, and epistemology) on the development of linguistics and employed Hülser to work on it.

Karlheinz Hülser (born 1942) was uniquely qualified for the job: he had come to Konstanz to do his doctoral work with Friedrich Kambartel on Wittgenstein's *Tractatus* (1977) while he was simultaneously employed as a student worker in Greek Studies. He stayed in Konstanz for his entire career as a lecturer in philosophy. It soon became clear, however, that the project required much more reliable source materials than were available at the time. Hülser thus started work to amend the situation, work which became the monumental, 2,000-pages edition of all then-known fragments of Stoic logic: *Die Fragmente zur Dialektik der Stoiker* (1987–88; four volumes). As an external reviewer (G. Nuchelmans) stated it: "Mr. Hülser deserves ample admiration and gratitude for the perseverance and acumen with which he has brought a truly daunting enterprise to an end that will be an indispensable starting-point for any future worker in this thorny field."

This was the heyday of logic in linguistics (or, to be more specific, in formal semantics) where faculty and students lived the interdisciplinary dream envisioned by the *Gründungsausschuss*. They spent much time in the centrally located canteen and interrupted their discussions only for short visits to a nearby classroom to use one of the chalkboards. Students and faculty alike sat in on one another's classes whether it was linguistics, philosophy, or mathematical logic. Linguists Heim and Zimmermann, for example, got to know each other in a philosophy class taught by Gottfried Gabriel, while philosophers Gabriel and Pirmin Stekeler-Weithofer frequented not only events hosted by linguists (e.g., the SFB 99) but also by literary studies, in particular the *Forschergruppe Poetik and Hermeneutik*, a significant national research cooperation that Hans Robert Jauß had brought to Konstanz.[2] Linguists had their research colloquium, featuring both internal and external speakers, every Thursday from 4 to 6pm, and philosophers had theirs immediately afterwards from 6 to 8pm, so faculty went to both. (Mathematicians met for their colloquium Fridays at 5pm.) Faculty and students would mingle at private parties, go on annual department-wide hiking trips in the Alps (a custom followed by both Linguistics and Philosophy), face each other once a year for a soccer match ('don't foul Mittelstraß,' linguists cautioned one another, 'otherwise you might kill philosophy of science in Germany'), or rent an entire pub for a common graduation party. Angelika

---

[2]Some may argue it influenced how Gabriel and Stekeler-Weithofer thought and wrote about logic.

Kratzer recalls her formative years in Konstanz with the words: "it was pure Utopia—something that wasn't available anywhere else in Germany (or in the world)."

**Internal cooperations.** While the informal exchange was thick and frequent, the depth of formal cooperations differed. Gerhard Neubauer (1930–2003), who worked in functional analysis (operator theory in Banach spaces), was the first mathematician to join the new university, and he took an active interest in the education of students and curricular offerings. In the spirit of the original idea of the *Gründungsausschuss* to organize mathematics as part of the *Interfakultät*, he took it upon himself to teach the service course "Mathematics for Linguistics" when no one else volunteered. It was mostly automata theory. This changed, and it changed dramatically, when Ulf Friedrichsdorf got appointed to a permanent position as lecturer in the Department of Mathematics. He not only taught "Mathematical-logical foundations for linguists" on a regular basis during his entire time at the university, sometimes team-taught with Egli or another linguist, but would actively advise linguistics faculty on their projects and lend them his full support as a trained mathematician and logician. During the long time both held faculty positions, Friedrichsdorf and Egli team-taught classes on a wide variety of topics at the intersection of logic and linguistics, and they did it almost every semester, for a while joined by von Heusinger, and sometimes with other linguists. Topics included type theory, formal semantics, formal languages, and non-monotonic reasoning. For more than three decades Friedrichsdorf served as the logic hinge between linguistics and mathematics.

The same cannot be said for philosophy, at least not without further qualifications. People knew each other very well and talked to one another frequently, but they rarely committed to a formal cooperation. And if they did, then it did not require interdisciplinary work in the more narrow sense of two or more researchers with different disciplinary affiliations teaming up. (One could argue that this was not required since the project leads had the double competency needed to execute their projects without cross-disciplinary cooperation.) For example, philosophers had individual projects within a *Sonderforschungsbereich* run by linguists (e.g., Kambartel had one in SFB 99 and Wolfgang Spohn in SFB 471), but those do not look like integral parts of the bigger project. This began to change when the philosopher-linguists Kamp (who took over from Zimmermann) and Ulrike Haas-Spohn had individual projects in the *Forschungsgruppe* "Logic in Philosophy," and turned into a true collaboration with the *Forschungsgruppe* "What if," where Konstanz linguists Maribel Romero, Riccardo Nicolosi, and Maria Biezma not only each had their own project but also worked closely with their counterparts in philosophy. In closing we should mention that Egli and Hubert Schleichert (philosophy) co-authored a bibliography on erotetic logic for a volume edited by Nuel D. Belnap.

**Selected publications.**
Bäuerle, R., Egli, U., and von Stechow, A. (eds). *Semantics from Different Points of View* (= Springer Series in Language and Communication; 6), Berlin: Springer (1979).

Bäuerle, R., Schwarze, Ch., and von Stechow, A. (eds). *Meaning, Use and Interpretation of Language* (= Foundations of Communication and Cognition 6), Berlin: De Gruyter (1983).

Brockhaus, K. and von Stechow, A. "On formal semantics: a new approach," in: *Linguistische Berichte*, 11 (1971), pp. 7–36.

Brockhaus, K. and von Stechow, A. "Mathematische Verfahren in der Linguistik," in: *Grundriß der Literatur- und Sprachwissenschaft, Vol. II: Sprachwissenschaft*, ed. H. L. Arnold and V. Sinemus, Munich: dtv (1973), pp. 61–90.

Cresswell, M. and von Stechow, A. "De re belief generalized," in: *Linguistics and Philosophy*, 5 (1982), pp. 503–535.

Egli, U., Pause, P., Schwarze, Ch., von Stechow, A., and Wienold, G. (eds). *Lexical Knowledge in the Organisation of Language* (= Current Issues in Linguistic Theory; 114), Amsterdam: Benjamins (1995).

Kratzer, A., Pause, E., and von Stechow, A. *Einführung in Theorie und Anwendung der generativen Syntax*, Frankfurt: Athenäum (1974).

Schepping, M. and von Stechow, A. (eds). *Fortschritte in der Semantik. Ergebnisse aus dem Sonderforschungsbereich 99 „Grammatik und sprachliche Prozesse"der Universität Konstanz*, Weinheim: VCH (1988).

von Stechow, A., and Wunderlich, D. (eds). *Semantik. Ein internationales Handbuch zeitgenössischer Forschung*, Berlin: De Gruyter (1991).[3]

## 3 Logic in philosophy

It is not entirely clear how the *Gründungsausschuss* arrived at the decisions they made. Joachim Ritter (1903–1974), for example, then a philosopher at the University of Münster, was a representative of traditional German philosophy and advocated for its traditional place in a university, while people around the eminent sociologist Ralf Dahrendorf (1929–2009) envisioned "eine Nicht-Hegelische Universität ... in der die Philosophische Fakultät im Hintergrund steht, wenn es sie überhaupt gibt"[4] and instead championed the empirical sciences which included a critical reflection on their own methodology. No matter the differences, the founding committee considered

---
[3] Many papers can be found in the journal *Linguistische Berichte* or were published in the pre-print series *Arbeitsberichte der Fachgruppe Sprachwissenschaft*, partially available online at the university's institutional repository.

[4] Translation: "a non-Hegelian university in which the humanities play a minor role or no role at all".

the mathematician-turned-philosopher Paul Lorenzen to be their first choice for filling a faculty line in philosophy. But some were afraid he would not be up to the administrative challenges, so another mathematician-turned-philosopher, but twenty years younger, was offered the job: Friedrich Kambartel. He came to Konstanz in 1968. This made him founding professor (*Gründungsprofessor*) and gave him some sway over the final design of the various schools and programs as well as a say in the filling of other faculty lines. Kambartel thus had a voice in the hire of Jürgen Mittelstraß in 1970, a voice both had when Peter Janich was recruited in 1973. Three faculty lines of the first generation had thus been filled with people who affiliated themselves with the Erlangen school of philosophy.

On one hand, this was good news as far as logic was concerned. It meant that logic would play a prominent role in philosophy from day one. After all, the book by Kamlah and Lorenzen, *Logical Propaedeutics. A Rational Speech Primer* (*Logische Propädeutik. Vorschule des vernünftigen Redens*, 1967, [2]1972)—widely considered the manifesto of the Erlangen School—had made quite a splash at the time. On the other hand, it spelled doom for some, for it was not clear whether logic in Konstanz would be in the shackles of an opinionated and idiosyncratic philosophy. This concern was somewhat unfounded and turned out to be true only to a certain and dwindling extent; traditional (i. e., non-formal) philosophical logic actually flourished.

In respect to modern, formal logic, the only viable alternative to the Erlangen School at the time was the Munich school of Wolfgang Stegmüller who taught logic and the philosophy of science like analytic philosophers did elsewhere in the world. And slowly Munich took over Konstanz. The initial agent of change was Peter Schroeder-Heister. He facilitated the hiring of André Fuhrmann, a product of St Andrews and Australian National University, and of Hans Rott, a former doctoral student of Stegmüller in Munich, but who had been advised by Wolfgang Spohn. Three logicians, none of them an Erlangen faithful, were now working in the department. Six years after Rott, in 1996, Spohn joined faculty ranks at Konstanz. And while he was more than a logician, he was also that. His first slate of hires added four more: Bernd Buldt, Volker Halbach, Holger Sturm, and Max Urchs. Seven logicians working in the same department (Schroeder-Heister had moved to nearby Tübingen): that was unprecedented. And even when their number fluctuated and eventually dropped in subsequent years, philosophical logic, both formal and non-formal, thrived for the next 25 years. And the department built on that strength when they hired Leon Horsten, more a logician than anything else, as Spohn's successor.

One more remark. The denomination of the three chairs in philosophy was reminiscent of the original idea, discussed by members of the *Gründungsausschuss*, not to have a stand-alone philosophy department but to make philosophy part of the *Interfakultät* and, consequently, to assign one chair to each of the three disciplinary clusters (i.e., natural sciences, social sciences, and humanities) and task their holders with the critical reflection of both methods and basic assumptions specific for the disciplines in that cluster. The denominations thus read "professor of philosophy, with special emphasis on the philosophy of the exact sciences," or "... the social sciences" and "... the humanities," respectively. These position were filled with Kambartel (in Konstanz: 1968–1993), Albrecht Wellmer (in Konstanz: 1974–1990), and Mittelstraß (in Konstanz: 1970–2005). Wellmer had habilitated with Jürgen Habermas in Frankfurt and was hence seen as a proponent of the Frankfurt School or Critical Theory. The Frankfurt School and the Erlangen School had a common target that united them at the time: an uncritical, affirmative positivism. To some extent, they were allies.[5] This gave philosophy in Konstanz, whose sole focus was on the philosophy of science, a coherence that was absent from any other philosophy department in Germany at the time where diversification, not concentration, was regarded paramount.

**Era Kambartel & Mittelstraß: 1970–2009**

**People: their stories & their projects.** Friedrich Kambartel (1935–2022) had earned a doctoral degree in mathematics (in complex analysis, *Funktionentheorie*, to be more specific) with Heinrich Behnke in Münster (1960) before he obtained his habilitation in philosophy (published in 1968 as *Erfahrung und Struktur*). He joined the University of Konstanz in 1968 but left for Frankfurt in 1993 (after Jürgen Habermas had convinced the state department of education to flout their own rules in order to make the job offer possible). In his work on topics like rationality or the foundations of modern science he emphasized the role of both practical reason and—possibly an influence of Ritter—culture: reason is not a free-floating entity but present only in its cultural manifestations.

Jürgen Mittelstraß (born 1935) completed both his doctoral work (1961, published 1962, *Die Rettung der Phänomene*) and his habilitation (1968, published 1970, *Neuzeit und Aufklärung*) in Erlangen, from where he came to Konstanz in 1970. He initiated, and became director of, the *Zentrum Philosophie und Wissenschaftstheorie* in 1990 and retired in 2005. Mittelstraß was stupendously productive and worked on many topics in the philosophy of science, broadly conceived (not to mention his extensive service as a government consultant), but one focal point that permeates his work is the Erlangen emphasis on *praxis*: science and its rationality is a certain way of

---

[5]Kambartel recalls not without a sense of pride that it was he and Mittelstraß who drafted and then promoted the 'Manifesto of the One-Hundred,' that is, the declaration against the Federal anti-radical decree (*Radikalenerlass*) signed by one hundred professors.

living (*Lebensform*), or doing things, which can then be used to argue for, defend, and justify it. But, going beyond Kamlah and Lorenzen, he pursued it in a post-Kuhnian way.

Peter Janich (1942–2016) obtained his doctoral degree with Paul Lorenzen (1969: *Protophysik der Zeit*). His research agenda was to continue his doctoral work and to provide all of the sciences with their respective protosciences; in other words, his goal was to identify a set of orthopractices—a collection of pre-scientific, artisan technical and measurement skills—that lend meaning to the basic vocabulary of the corresponding science and would thus define and justify in a non-circular way the fundamental concepts of that science. This made him the most hard-nosed representative of the Erlangen School in Konstanz when he started in 1973.

All three shared a normative orientation, which led to some tensions when they attempted to evangelize students and colleagues and preach the gospel of Erlangen constructivism. But Kambartel and Mittelstraß grew mild(er) over the years, and Janich left in 1980 which meant that, by and large, efforts to use formal methods left with him. So we skip the further development. But all three were strong supporters of logic, and Kambartel and Mittelstraß, who stayed on, fostered logic, each in their own but complementary ways, as we will show now.

Kambartel brought important editorial projects to Konstanz: Frege's posthumous works, Bolzano's *Collected Works* as well as the subject editorship for logic, philosophy of language, analytic philosophy, and the history and philosophy of science of the magistral "Historical Dictionary of Philosophy" (*Historisches Wörterbuch der Philosophie*, 1971–2007, in 12+1 volumes, on 8,736 pages or 17,144 columns). It is curious that it was Ritter in Münster who encouraged Kambartel to pick up the slack and continue work on Frege's *Nachlass*, which lay abandoned since Scholz's death in 1956. Kambartel *et al.* published Frege's 'Posthumous Writings' in 1969 ($^2$1983) and his 'Correspondence' in 1976. His long-term collaborator, Gabriel, would bring two of the historical projects to a good ending: more of Frege's posthumous works (diary, lecture notes) as well the Historical Dictionary.

Kambartel had brought along his student Gottfried Gabriel (born 1943) in order to receive his continued assistance with the edition of Frege's posthumous works. Gabriel earned his PhD (it was the third doctoral degree conferred in Philosophy) with an essay on definition in 1972 (published as *Definitionen und Interessen*), while his habilitation thesis in 1975 was on the semantics of fictional speech (*Fiktion und Wahrheit*, 1975, rev. $^2$2019). It was an outcome of his broad interests that included not only Frege (cue: *ungerade Rede*) but also what was done next door where the program in literary studies was running the special interest group *Poetik und Hermeneutik*. Gabriel continued to be a voice in the history of logic; e.g., he edited and re-issued Lotze's *Logik* and worked on early Analytic Philosophy. He stayed in Konstanz until 1992 and then went first to Bochum and three years later to Jena.

Another project in the history of logic, while close to Kambartel, came out of linguistics and was mentioned already: Hülser's *Die Fragmente zur Dialektik der Stoiker*. While it soon became clear that there were obvious analogies between Stoic logic and Frege—Hülser recalls that he convinced Kambartel and Gabriel during hallway discussions in the winter semester 1981/82—it took three more decades to find the missing links and build a compelling case that some of Frege's views might indeed have been informed by Stoic ideas (see *History and Philosophy of Logic* 30:4 (2009), 369ff.).

What also deserves mention in the context of the history of logic is the *Philosophisches Archiv*. It started in 1978 as a modest effort by Gereon Wolters to collect historically important papers from Hugo Dingler's widow. The reason was that Dingler (1881–1954) was considered a forerunner of Erlangen ideas on proto-science. More acquisitions followed, skillfully negotiated by its now director, Wolters; and with institutional support by Mittelstraß, the *Dingler-Archiv* became the *Philosophisches Archiv* in 1985.

Gereon Wolters (born 1944) did his doctoral work on the axiomatic method in Johann Heinrich Lambert (1977) and his habilitation thesis on Ernst Mach (1985), both advised or promoted by Mittelstraß. Wolters became a tenured professor in Konstanz, working as a philosopher of science (particularly biology) and serving as the long-term director of the *Archiv* even beyond his retirement in 2009.

Facilitated by the personal relationships Mittelstraß had established with faculty at the Center for the Philosophy of Science in Pittsburgh (he had declined UPitt's offer in 1975) and his richly flowing grant money (among others, he was awarded the Leibniz Prize in 1989), a formal cooperation agreement was signed to the effect that the *Philosophisches Archiv* and the Archives for Scientific Philosophy in Pittsburgh share digital copies of their respective archival holdings. European researchers have thus gained access to the literary estates of eminent scholars like Oskar Becker, Rudolf Carnap, Bruno de Finetti, Kurt Gödel, Georg Kreisel, Paul Lorenzen, Frank P. Ramsey, or Ludwig Wittgenstein, among others.

Mittelstraß had been a student of Kamlah in Erlangen, not of Lorenzen, but the importance of logic was nevertheless part of his academic DNA. This shows clearly in the "Encyclopedia of Philosophy and Philosophy of Science" (*Enzyklopädie Philosophie und Wissenschaftstheorie*) which he shepherded through two editions (1980–1996, in four vols; 2nd edition, 2005–2018, in eight vols) and which covers logic to such an extent that it doubles as a comprehensive logic dictionary. And the fact that it includes person articles makes it even more useful, since logicians, from Aristotle to Zermelo, each have their own entry.

But Mittelstraß was also instrumental in bringing contemporary philosophical logic to Konstanz. In 1988 he launched, with ample funding by the state of Baden-Württemberg, the Center for Philosophy and Philosophy of Science (*Zentrum Philosophie und Wissenschaftstheorie*) as a campus-wide unit whose mission was to continue and revivify the idea of the *Gründungsausschuss*, namely, the idea of philosophy being integrated into the various sciences in its role of critically reflecting on their foundations and methodology. Peter Schroeder-Heister, first hired by Wolters to assist with the *Dingler-Archiv* and who in the meantime had become a close collaborator of Mittelstraß, recommended to include new or emerging fields such as computer science and artificial intelligence to the description of the Center's profile and advised him to open new faculty lines to applicants beyond the Erlangen school; Mittelstraß did both. This brought not only Martin Carrier to Konstanz, but also André Fuhrmann and Hans Rott; Schroeder-Heister thus doubled the number of formal logicians before he left.

Peter Schroeder-Heister (born 1953) had studied for a teaching degree (*Staatsexamen*) in mathematics and philosophy at the University of Bonn (1977) and was hired a year later by Wolters to work on the Dingler project. While in Konstanz, he was pursuing his doctoral work on proof theory, later accepted by Gisbert Hasenjäger in Bonn and—due to the most helpful intervention of Prestel—Dag Prawitz in Stockholm as external reader: *Untersuchungen zur regellogischen Deutung von Aussagenverknüpfungen* (1981). He worked in Konstanz on various logic-related topics, both historical and proof-theoretic, and obtained his habilitation in 1988, before he left for Tübingen in 1989 and became the face of proof-theoretic semantics.

Hans Rott (born 1959) earned his doctorate in Logic and Philosophy of Science in Munich, officially with Wolfgang Stegmüller but actually advised by Wolfgang Spohn: *Reduktion und Revision. Aspekte des nichtmonotonen Theorienwandels* (1989) and worked for a year with Kamp in Stuttgart, before Mittelstraß offered him a position in Konstanz (1990–1997). His habilitation thesis was *Making Up One's Mind. Foundations, Coherence, Nonmonotonicity* (1997). Soon after he joined the University of Amsterdam (1997–1999) but moved back to Germany and has been at the University of Regensburg since 1999. Formal theories of belief change was the topic he concerned himself with mostly during his Konstanz years.

André Fuhrmann (born 1958) had completed his MPhil at St. Andrews (1984) and his PhD at the Australian National University, *Relevant Logics, Modal Logics, and Theory Change* (1988), with Richard Sylvan (Routley), J. J. C. Smart, and Neil Tennant as his committee, before Schroeder-Heister brought him to Konstanz in 1989, where he obtained his habilitation with an *An Essay on Contraction* in 1995 and stayed on as lecturer and Heisenberg Fellow until 2002. He went to São Paulo (2002–2006) and has been in Frankfurt since 2006. During his time in Konstanz his research was mostly on non-classical logics and formal theories of belief change. He and Rott cooperated closely on the latter topic and were at the cutting edge of the field, recognized internationally for their contributions. They were joint recipients of the Heinz Maier-Leibnitz Prize in 1996.

Another cooperation came about by sheer serendipity. Luc Bovens had come to Konstanz on a Humboldt fellowship for the academic year 1998/99 with the plan to immerse himself in probability theory. During that year, he made some contact with Spohn and his group (see the section on the *Era Spohn*) but was assigned an office in the Mittelstraß area, namely, the office of his host, Fuhrmann, who was on sabbatical. This led to a chance encounter with his next door neighbor, Stephan Hartmann. Bovens inquired about Hartmann's interests and, upon learning about the other's background, he asked him for help to work through an introductory text on Bayesian networks. Hartmann, more interested in naturalized philosophy of science at the time, reluctantly agreed, and they ended up meeting regularly to study the theory of Bayesian networks and to further explore their potential. The real collaboration, however, got going only after Bovens had left Konstanz. It led to an avalanche of co-authored papers, a co-directed research group (PPM, see §5.1), and catapulted the two to the forefront of contemporary Bayesianism.

Stephan Hartmann (born 1968) studied physics and philosophy in Giessen (MS Physics, 1991; PhD in Philosophy in 1995 with Bernulf Kanitscheider: *Metaphysics and Method*). Concurrently with his doctoral work in philosophy, he was pursuing a PhD in physics (with Jürgen Audretsch, Konstanz, and later with Axel Schenzle, Ludwig-Maximilians-Universität München, LMU) when Mittelstraß hired him as an assistant professor in 1998. He left Konstanz for a position at the London School of Economics (LSE) in 2004 and moved to Tilburg in 2007 (where he founded the Tilburg Center for Logic and Philosophy of Science, TiLPS)). Since 2012 he has been professor of philosophy of science at LMU and co-director of the Munich Center for Mathematical Philosophy (MCMP), which Hannes Leitgeb founded in 2010.

Finally, we should mention Pirmin Stekeler-Weithofer (born 1952), since 1992 professor in Leipzig. He was considered something like a whiz kid, moving freely and competently between mathematics, philosophy, and linguistics, and did his initial work on Erlangen home turf: logic and mathematics. This brought him into conflict with his teachers; Kambartel, whose protégé he was, called him a "doubtful case" (*unsicherer Kandidat*), and Janich actually tried to obstruct his habilitation. As doctoral work, Stekeler-Weithofer conducted a critical investigation into key components of a formal logic (concepts, truth-functors, inference rules) whose results clashed with the game-theoretic Erlangen orthodoxy (1984, published 1986 as *Systeme der Logik. Eine Kritik der formalen Vernunft*), while in his habilitation thesis he developed his own take on a philosophy of mathematics (1987, published 2008 as *Formen der Anschauung. Eine Philosophie der Mathematik*), where he defended the primacy of geometry as a basis for mathematical thinking but severely criticized the proto-geometric program of the Erlangen School.

**Teaching, research, students.** Past schedules of classes do not seem to bear it out, but institutional memory has it that the logic education had been in the hands of Schleichert. Hubert Schleichert (1935–2020) had obtained both his doctoral degree (1957) and habilitation (1968) with Béla Juhos at the University of Vienna.[6] Schleichert became a professor in 1973 and was considered the logical positivist in the department.[7]

The course "Logical Propaedeutics," initially taught according to the eponymous book by Kamlah and Lorenzen, was mandatory for all majors and was followed by one or two courses called "Formal Logic." Typically, instructors would mention results like completeness but not prove them in class.[8] In addition to these introductory classes, Kambartel, Gabriel, and Hülser, quite regularly offered classes on Frege and other topics in history of logic, and occasionally other faculty (e.g., Mainzer) did as well. The Erlangen version of a game-theoretic justification of logical rules obviously played some role—e.g., Kambartel lectured on it—but it took a back seat as time went by. An exception was a brief teaching stint by Gerrit Haas who held a limited-term position after his MA with Lorenzen and before he started doctoral work with Christian Thiel in Aachen. He was excited about mathematical logic and metamathematics, presented full proofs in class, and was able to spread his enthusiasm; Stekeler-Weithofer recalls him as someone who inspired him.

Most research done by the people close to Mittelstraß and Kambartel was not related to logic, and to the extent it was, we briefly mentioned it in the preceding section as part of the thumbnail biographies.

Dissertations advised by the Kambartel-Mittelstraß circle which fall into the scope of the DVMLG include (here we list topics rather that titles and

---

[6] Juhos was a student of Schlick and, with Victor Kraft, what was left of the Vienna Circle in post-war Vienna. Juhos was also very briefly, from 1 April 1971 to his death on 27 May 1971, a member of the DVMLG; cf. B. Löwe, *Die Mitgliederentwicklung in der Frühzeit der DVMLG*, in this volume.

[7] This was not without irony since his position was for the history of philosophy; so he offered lecture courses on Nietzsche that drew big crowds from across campus but also taught Chinese Philosophy.

[8] This did not really change until Rott and Fuhrmann, and then Spohn's people, taught these (and other) classes.

do not repeat those we already mentioned above): Carlos Pereda (argumentation, 1974), Wolfgang Kemnitz (subjective probability, 1976), Karlheinz Hülser (early Wittgenstein, 1977), Peter Georgi (Greek mathematics, 1989), Alexander Rüger (quantum field theory, 1989), Fernando Augusto da Rocha Rodrigues (propositions and objects, 1990), Ursula Klein (emergence of chemistry, 1993), Edgar da Rocha Marques (Wittgenstein, 1995), Wolfgang Kienzler (late Wittgenstein, 1995), Mechthild Jäger (constructivism, 1997), Ulrich-Ekkehard Sauter (quantum mechanics, 1998), Veiko Palge (quantum mechanics, 2006). Wellmer served as primary advisor of Maeve Cooke (formal pragmatics, 1989), and two more dissertations in the philosophy of the exact sciences were advised by Paul Hoyningen-Huene: Marcel Weber (theory of evolution, 1996) and Insok Ko (thermodynamics, 1997).

**Notable projects, conferences, & guests.** We noted already major editorial projects—e.g., Frege's posthumous works or the *Enzyklopädie Philosophie und Wissenschaftstheorie*—as well as significant institutional projects: the *Philosophical Archive* or the *Zentrum Philosophie und Wissenschaftstheorie*. But the cooperation between Konstanz and Pittsburgh did not only bolster archival holdings at both institutions (see above) but also furthered the exchange among philosophers of science in the two countries via a series of conferences, the so-called "Pittsburgh–Konstanz Colloquium in the Philosophy of Science." The Conference met every two years, alternated between Konstanz and Pittsburgh, and the proceedings were published with the University of Pittsburgh Press; it fizzled out when those whose personal friendships had sustained it retired. Other international conferences were in the area of research by Fuhrmann and Rott: "The Logic of Theory Change" (October, 1989) organized by Fuhrmann and Michael Morreau (Tübingen, now Tromsø); "LogIn—Konstanz Colloquium in Logic and Information" (October, 1992), organized by Fuhrmann and Rott. Rott and Sven Ove Hansson (Uppsala, now Stockholm) coordinated a joint research project, funded by the DAAD, "Wissensrevision / Belief revision" (1993–1996), that involved 13 researchers from the universities of Konstanz, Leipzig, Lund, Saarbrücken, Umeå, and Uppsala.

Obviously, the Mittelstraß group saw many international guests and visitors; nearly everyone who had a name in analytic philosophy or the philosophy of science passed through Konstanz at least once to give a talk. A special role, however, was played by Fuhrmann. He capitalized on his education at international centers of logical research, groomed his professional network, and brought many scholars to Konstanz for an extended stay. This is how Luc Bovens, for example, came to Konstanz; Fuhrmann had met him at the annual Czech workshop *Logica*. Other names that may ring a bell among members of the DVMLG and who came for extended stays include George Boolos (MIT), Gabriella Crocco (Paris, now Marseille), Mic Detlefsen (Notre Dame), Michael Friedman (Indiana U, now Stanford), Kosta Došen (Belgrade), Andreas Herzig (Toulouse), Jean-Pierre Marquis (Montreal), Ingolf Max (Leipzig), David McCarty (Indiana U), Kazuyuki Nomoto (Tokyo), Francesco Paoli (Cagliari), Jaroslav Peregrin (Prague), Uwe Scheffler (Dresden), Tomasz Skura (Zielona Góra), Igor Urbas (Canberra, then Madrid), and Max Urchs (Leipzig). It seems the Mittelstraß group did what they could to help logicians from East Germany to weather the changes the German reunification had brought about.

**Selected publications.**

Bovens, L. and Hartmann, S. "Bayesian networks and the problem of unreliable instruments," in: *Philosophy of Science*, 69 (2002), pp. 29–72.

Bovens, L. and Hartmann, S. "Solving the Riddle of Coherence," in: *Mind*, 112 (2003), pp. 601–634.

Došen, K. and Schroeder-Heister, P. "Uniqueness, definability and interpolation," in: *Journal of Symbolic Logic*, 53 (1988), pp. 554–570.

Fuhrmann, A. "Theory contraction through base contraction," in: *Journal of Philosophical Logic*, 20:2 (1991), pp. 75–203.

Fuhrmann, A. "When hyperpropositions meet," in: *Journal of Philosophical Logic*, 28:6 (1999), pp. 559–574.

Fuhrmann, A. "A relevant theory of conditionals," in: *Journal of Philosophical Logic*, 24:6 (1995), pp. 645–665.

Fuhrmann, A. and Morreau, M. (eds). *The Logic of Theory Change* (= Lecture Notes in Artificial Intelligence; 465), Berlin: Springer (1991).

Fuhrmann, A. and Rott, H. (eds). *Logic, Action, and Information. Essays on Logic in Philosophy and Artificial Intelligence*, Berlin: De Gruyter (1996).

Gärdenfors, P. and Rott, H. "Belief revision," in: *Handbook of Logic in Artificial Intelligence and Logic Programming. Vol. 6: Epistemic and Temporal Reasoning*, ed. D. M. Gabbay, C. J. Hogger, J. A. Robinson, Oxford: Oxford UP (1995), pp. 35–132.

Gethmann-Siefert, A. and Mittelstraß, J. (eds). *Die Philosophie und die Wissenschaften. Zum Werk Oskar Beckers*, München: Fink (2002).

Heinzmann, G. and Wolters, G. (eds). *Paul Lorenzen – Mathematician and Logician* (= Logic, Epistemology, and the Unity of Science; 51), Cham: Springer (2021).

Mittelstraß, J. (ed.). *Paul Lorenzen und die konstruktive Philosophie*, Münster: Mentis (2016).

Mittelstraß, J. and von Bülow, Ch. (eds). *Dialogische Logik*, Münster: Mentis (2015).

Rott, H. "Belief contraction in the context of the general theory of rational choice," in: *Journal of Symbolic Logic*, 58 (1993), pp. 1426–1450.

Rott, H. "Preferential belief change using generalized epistemic entrenchment," in: *Journal of Logic, Language and Information*, 1:1 (1992), pp. 45–78.

Rott, H. *Change, Choice and Inference: A Study of Belief Revision and Nonmonotonic Reasoning* (= Oxford Logic Guides; 41), Oxford: Oxford UP (2001).

Schroeder-Heister, P. "The completeness of intuitionistic logic with respect to a validity concept based on an inversion principle," in: *Journal of Philosophical Logic*, 12 (1983), pp. 359–377.

Schroeder-Heister, P. "Popper's theory of deductive inference and the concept of a logical constant," in: *History and Philosophy of Logic*, 5 (1984), pp. 79–110.

Schroeder-Heister, P. "A natural extension of natural deduction," in: *Journal of Symbolic Logic*, 49 (1984), pp. 1284-1300.

## Era Spohn: 1996–2018

**People: their stories & their projects.** As far as logic is concerned, the hiring of Wolfgang Spohn was considered a game changer. This is curious for the following reason. Spohn hailed from Munich, where he did doctoral work on the conceptual foundations of decision theory (Munich 1976, published in 1978 as *Grundlagen der Entscheidungstheorie*) and developed a theory of causality in his habilitation (Munich, 1983), which already contained the foundations of a new, non-classical theory of probability (the theory of ranking functions), developed to better suit philosophical needs (e.g., when it comes to modeling the dynamics of belief change). Thus, when he joined the department, he came as a recognized expert in decision and probability theory and their ramifications for philosophy, and in particular the philosophy of science and formal epistemology, and few will have known that his unpublished master's thesis had been devoted to deontic logic, of which a single published paper bore witness compared to the roughly two dozen articles he had published in the former field before he came to Konstanz. But, being a student of Wolfgang Stegmüller and seen by many as his rightful heir, he was considered a logician. And he did not disappoint. While he continued research along his own lines, he promoted and supported logic in all its shapes and forms, whether historical, philosophical, or mathematical. Moreover, logical research now had a mentor who himself was very well-versed in its mathematics as well as in its philosophical aspects. Finally, he was able to direct research also in the cognate fields of formal linguistics, where he had a peer in his wife, Ulrike Haas-Spohn, who was a philosopher-linguist and had studied, among others, with von Stechow.

But this was not the only change Spohn's arrival in Konstanz ushered in. First, Kambartel and Mittelstraß framed rationality and logic as resulting from and being embedded in cultural practices, with language being one among other practices. Broad historical considerations therefore played a prominent role. For Spohn, however, the medium in which rationality expresses itself was just language; not the language spoken by warm bodies enmeshed in a specific culture, mind you, but the de-historized stripped-down—'sanitized,' if you will—language of formal linguists and analytic philosophers. Regardless of whether one embraces or rejects such a reduction in complexity, it is a prerequisite for bringing the tools of mathematics and formal logic to the task and thus helps explain the increase in formal studies during Spohn's tenure. Second, Konstanz–Erlangen philosophy of science was mostly a national discourse; parochial in its decision to publish the majority of its books and articles in German. Analytic philosophy represented by Spohn was international and spoke English. Last but not least, funding went to a lesser degree to individuals to bolster their institutional standing but rather to support teamwork and cooperation. All in all, it was a different style of doing philosophy.

From his previous position in Bielefeld, Spohn brought along Bernd Buldt (in Konstanz: 1996–2002 & 2004–2006)—who had done his doctoral work on Gödel (*Die Sätze von Gödel. Logische und philosophische Perspektiven*, Bochum, 1991, supervised by Gert König and with Jürgen von Kempski, one of the founding members of the DVMLG, in the wings) but shifted his attention to the philosophy of mathematics and the history of probabilistic reasoning during his years in Konstanz—and filled his second line with Volker Halbach (in Konstanz: 1997–2004) who, building on his dissertation (*Tarski-Hierarchien*, Munich 1994), established himself as a recognized authority on formal theories of truth during his time in Konstanz. They were soon joined by Sturm and Urchs. Holger Sturm (in Konstanz: 1997–1998 & 2003–2013), who had worked on infinitary polymodal logics for his PhD (*Modale Fragmente von $\mathscr{L}_{\omega\omega}$ und $\mathscr{L}_{\omega_1\omega}$*, Munich, 1997), moved from research in computer science logics via a study of properties to a general account of meaning, while Max Urchs (in Konstanz: 1998–2005)—whose concurrent job at the University of Szczecin (Poland) could not support a family—continued work on non-classical logics (Jaśkowski systems of deduction, temporal and causal logic), work he had begun in his PhD ("Systems with J-Implications Based on Multidimensional Modal Calculi," Copernicus University, Toruń, 1982) and his habilitation ("Causal Logic," Leipzig, 1987).

Spohn's initial slate of hires were thus all logicians. Other junior researchers Spohn put on faculty lines were not primarily logicians but reflect the wide spectrum of analytic philosophy that he stood for and whose work still falls in the broad range of topics covered by the DVMLG: Ludwig Fahrbach (in Konstanz: 2001–2003) worked on Bayesianism, Wolfgang Freitag (in Konstanz: 2005–2012) on epistemology, Gordian Haas (in Konstanz: 2001–2003) on belief revision, Manfred Kupffer (in Konstanz: 1997–1998 & 2003–2004) on counterparts (i. e., Lewisian metaphysics), and Tobias Henschen (in Konstanz: 2013–2018) on causality (in economics). The exception to these later hires is Alexandra Zinke (in Konstanz: 2013–2017). She was not only the only woman but worked on logic more narrowly defined and wrote her dissertation on the concept of logical consequence, work she continued later.

Beyond faculty lines tied to Spohn's endowment as chair, there were also a number of affiliated positions and/or researchers funded by research grants. Those who stayed for an extended period of time include Olsson, Merin, and Raidl.

Erik J. Olsson (in Konstanz: 1997–2003), had done his doctoral work on belief change with Sven Ove Hansson in Uppsala (1997), was part of the research group "Logic in Philosophy" (see page 73), and worked mostly on coherence and belief revision. After very productive years in Konstanz, he returned to Sweden (Uppsala, 2001; Lund, 2007).

Arthur Merin (in Konstanz: 1999–2014), who had done his PhD in Cambridge and died prematurely in 2019, straddled the borders between logic, decision theory, and formal semantics in various projects over a 15-year period.

Eric Raidl (in Konstanz: 2012–2018) worked on two projects: first as a member of the French-German cooperation "Causality and Probability", then as part of the research group "What if?" (see pages 74). He wrote his dissertation supervised by Jacques Dubucs and Spohn at Paris I–Sorbonne, *Probabilité, Invariance et Objectivité* (2014). He completed his habilitation *Conditional(s)* in Konstanz (2021) and started working on epistemology and the logics of Machine Learning in Tübingen (2019).

Early career faculty with shorter employment include Christoph Fehige, Manfred Kupffer, Franz Huber, Michael Baumgartner, Luke Fenton-Glynn, and Niels Olsen; their projects are mentioned below. Frank Zenker (in Konstanz: 2015–2017) had written a dissertation on belief revision (with Ulrich Gähde in Hamburg, 2007, and Olsson as external reader) and worked on his own project: "Conceptual Spaces, Reasoning, and Argumentation," funded by the Volkswagen Foundation. In addition to Zenker, Spohn also served as a host for two fellows at the *Zukunftskolleg*: Huber and Antos (see §§ 5.2 & 5.5).

Many of those who were hired by Spohn or were in his orbit obtained permanent faculty positions later: Baumgartner (Bergen, 2017), Buldt (Purdue, 2006), Fehige (Saarbrücken, 2008), Fenton-Glynn (UCL, 2013), Freitag

(Freiburg, 2012, then Mannheim, 2018), Haas (Bayreuth, 2011), Halbach (Oxford, 2004), Huber (Toronto, 2013), Olsson (Uppsala, 2001, then Lund, 2007), Sturm (Saarbrücken, 2013), Urchs (EBS Business School, 2006), Zenker (Lund, 2014), and Zinke (Frankfurt, 2022).

**Teaching, research, students.** Spohn and his collaborators offered the usual slate of lecture classes and seminars in analytic philosophy and continued teaching courses that were already on the books: "Logical Propaedeutics" and "Formal Logic I+II." What they added were introductions to set theory, mathematical logic, recursion theory, proof theory, provability logic and other modal or non-classical logics, formal theories of truth, game theory, and probability theory, all of which, while pegged towards an audience of philosophers, were taught to more rigorous standards and with an emphasis on proof. Classes on more advanced topics such as Rosser sentences, admissible sets and structures, or paradoxes were offered for a mixed audience and team-taught with Friedrichsdorf. Over many semesters, Buldt, Halbach, and Friedrichsdorf ran a team-taught seminar on various topics at the intersection of logic and the philosophy of mathematics, while Sturm and Friedrichsdorf did the same later when they cooperated on a theory of properties.

Spohn was fluent in all of analytic philosophy, so there was little that was not represented by some graduate or post-graduate student affiliated with his chair. Areas of concentration that fall into the scope of the DVMLG were four clusters: (i) probability theory (Bayesianism, belief revision, ranking functions); (ii) philosophy of the exact sciences (causality, philosophy of mathematics, philosophy of physics); (iii) logic (modal and non-classical logics, theories of truth and meaning); (iv) philosophy of language (formal semantics, counterfactuals). It is fair to say (I hope) that Konstanz gained a national, if not international reputation, for research in all four areas during Spohn's tenure as chair. He was given the Lakatos Award in 2012 and awarded the Frege Prize in 2015 for a reason. We move a more detailed account of these four clusters to the section on notable projects.

Dissertation and habilitation theses that fall into the scope of the DVMLG include (we do not reproduce exact titles but list topics): Lucas Amiras (protogeometry, 1999), Manfred Kupffer (counterparts, 2000), Ludwig Fahrbach (Bayesianism, 2000), Marion Ledwig (Newcomb's problem, 2000), Jacob Rosenthal (probabilities as propensities, 2002), Radu Dudau (realism/antirealism, 2002), Gordian Haas (theory change, 2003), Wolfgang Freitag (formal philosophy, 2005), Stefano Bigliardi (ranking functions, 2008), Stefan Hohenadel (belief networks, 2012), Benjamin Bewersdorf (belief revision, 2012), Alexandra Zinke (logical consequence, 2013), Robert Michels (modality, Geneva, 2013), Niels Skorgaard-Olsen (ranking theory, 2014), Eric Raidl (probability, 2014), Anna-Maria Eder (rationality, 2018),

Ali Zolfagharian (suspending judgment, 2020), Arno Goebel (if-constructions, 2020); Christopher von Bülow (structuralism, tbd); habilitations include: Michael Esfeld (holism in quantum mechanics, 2000), Volker Halbach (deflationism, 2001), Bernd Buldt (19th century probability, 2003), Erik J. Olsson (coherence, 2003), Wolfgang Freitag (theory of knowledge, 2010), Holger Sturm (theories of meaning, 2010), Gordian Haas (verificationism, Bayreuth, 2011), Tobias Henschen (causality in macroeconomics, 2017), and Eric Raidl (conditionals, 2021).

**Notable projects, conferences, & guests.** The first in a series of major collaborative research efforts that characterized Spohn's time as chair was the *Forschungsgruppe* (research group) "Logic in Philosophy" (1997–2003). It was Schroeder-Heister in Tübingen who had first conceived of it and had made preliminary plans, but when Spohn came to nearby Konstanz, they joined forces; the philosopher-linguist Kamp entered from Stuttgart at halftime, taking over from Zimmermann. The research group was funded by the DFG, and it was the first of its kind in philosophy, blazing a trail for how to conduct philosophy in a cooperative style. It ran for a total of six years (1997–2003) with a third-year review when some projects were renewed, some were not, while a few new ones were added. The initial subdivision and their individual projects were as follows (we list the PI after the semicolon):

(1) Logic and Epistemology:

   (i) Believing as deciding (Konstanz: Fuhrmann, Olsson; Rott);

   (ii) Coherence theories of knowledge (Konstanz: Fahrbach, Halbach; Spohn);

   (iii) Logical form of belief ascription (Stuttgart-Tübingen: Robert van Rooij, later: Haas-Spohn; Zimmermann)

(2) Logic and Metaphysics:

   (i) Necessity, logic, individuation (Konstanz: Wolfgang Benkewitz; Spohn)

(3) Basic Logical Concepts:

   (i) Truth and reflection (Tübingen: Walter Hoering);

   (ii) Proof-theoretic semantics (Tübingen: Reinhard Kahle, Patrizio Contu; Schroeder-Heister).

The three projects that failed to attract funding (by Buldt, Friedrichsdorf-Fuhrmann, and Hoering) were too mathematical, it seems. Projects that were added at halftime were: (i) Cognitive and referential aspects of concepts (Stuttgart-Konstanz: Haas-Spohn; Kamp); (ii) The semantic conception of the a priori (Stuttgart-Konstanz: Kupffer; Kamp, Spohn); (iii) a structural theory of properties (Konstanz: Sturm; Friedrichsdorf). The group produced 120 preprints, and about the same number of publications; moreover, group members organized about 15 workshops and conferences, twelve of which met in Konstanz.

Equaling in size, but, as members moved to other places, distributed over more locations than the first *Forschungsgruppe*, was the DFG-funded research group "What if?" (2012–2019). Its first phase was concerned with the epistemology and scientific relevance of counterfactual statements and thought experiments (2012–2015), while its second phase also included pragmatic, psychological, and cultural aspects of counterfactual thinking (2015–2019). It was mostly a collaboration among linguists (Maria Biezma, Riccardo Nicolosi (München), Maribel Romero) and philosophers (Bernhard Kleeberg (Erfurt), Thomas Müller, Tobias Rosefeldt (Berlin), Spohn, Marcel Weber (Geneva), Paul Ziche (Utrecht)) and one psychologist (Eva Rafetseder (Stirling)). It comprised the following individual projects that fall into DVMLG territory (we paraphrase some of the titles): (i) Conditionality, counterfactuality, and information transfer (Eric Raidl, Merin; Spohn); (ii) Semantics and pragmatics of counterfactuals (Eva Csipak, David Krassnig, Brian Leahy, Andreas Walker; Romero); (iii) Counterfactual knowledge and imagination (Daniel Dohr; Rosefeldt); (iv) Counterfactual thought experiments in the sciences (Julian Bauer; Kleeberg); (v) Alternatives for the future (Hadil Karawani, Antje Rumberg; Müller); (vi) Conditionals in discourse (Biezma); (vii) Instituting and contesting scientific openness (Ziche; Spohn); (viii) Simulation in neuroscience (Weber). The research group published close to 50 peer-reviewed articles and organized about 25 conferences and workshop, half of which met in Konstanz.

Research cooperations at a smaller scale or with less DVMLG-related topics include the following. The *Sonderforschungsbereich Entwicklung und Variation im Lexikon* (1997–2008), was another one that came out of linguistics, this time with Aditi Lahiri as speaker (Frans Plank took over when she left for Oxford in 2007), and with 41 individual projects it was a massive undertaking. Spohn participated with a project to bring decision-theoretic semantics to bear on questions of conceptual content (Benkewitz, Merin; Spohn). It was a continuation of a project, funded by the Thyssen Foundation, on the relevance theory of meaning (1999–2001, Merin; Spohn). The French-German cooperation "Causality and Probability" (2009–2012), co-sponsored by the ANR and DFG, was co-led by Jacques Dubucs (Paris). Participating researchers were Anouk Barberousse, Isabelle Drouet, Philippe Huneman, and Max Kistler in Paris as well as Michael Baumgartner, Lorenzo Casini, Luke Glynn, and Eric Raidl in Konstanz. Topics of investigation were (i) actual causation, (ii) counterfactual accounts of causation,

(iii) multi-level causation, and (iv) the objective reality of causation. The cooperation produced close to 40 peer-reviewed publications and organized five conferences or workshops, two of which met at Konstanz. The two projects "Reflexive Rationality: A Theory of Dynamic Choice," and "Reason Relations, Argumentation, and Conditionals: Applying Ranking Theory to Psychology of Reasoning" (2014–2017) were part of the first and second phases, respectively, of the national *DFG Schwerpunktprogramm* "New Frameworks of Rationality" (2011–2019), whose speaker was Markus Knauff (Gießen). Niels Skovgaard-Olsen collaborated with Spohn on both projects. A late outcome of this national cooperation was the *Handbook of Rationality* (2021), edited by Knauff and Spohn, which attempts to be, with 65 chapters in 15 sections on almost 1,000 pages, alarmingly comprehensive. Still ongoing is "Reflexive Decision and Game Theory" (2020–2025), a DFG-funded Koselleck Project. Its goal is to re-evaluate and hopefully transform the very basis on which contemporary game and decision theory have been erected, especially in respect to their normative role (what is rational behavior?). Current post-docs are Gerard Rothfus (PhD, Irvine, 2020) and Mantas Radzvilas (PhD, LSE, 2016), while İrem Portakal (TU Munich) and Bernd Sturmfels (MPI Leipzig) serve as associate members.

Projects and conferences from within Spohn's group but without his direct involvement include "Truth, Necessity and Provability" (1999), an international workshop in Leuven organized by Leon Horsten and Halbach; "Philosophy of Mathematics: Sociological Aspects and Mathematical Practice" (2006-2010), a DFG-funded international research network, initiated by Benedikt Löwe and Thomas Müller, with Buldt as a founding member; "Rudolf Carnap" (2006) an international workshop at GAP.6, organized by Steve Awodey and Buldt; "Towards a New Epistemology of Mathematics" (2006), an international workshop at GAP.6, organized by Buldt, Löwe, and Müller.

The Spohn group saw many international guests at their conferences, but not so many stayed for an extended period of time; those who did include Steve Awodey (Carnegie Mellon University), André Carus (Chicago, now Munich), and David McCarty (Indiana U).

**Selected publications.**
Baumgartner, M. and Glynn, L. (eds). *Actual causation*, Suppl. 1 to *Erkenntnis*, 78 (2013).

Buldt, B., Halbach, V., and Kahle, R. (eds). *Reflections on Frege and Hilbert*, special issue of *Synthese*, 147:1 (2005).

Buldt, B., Löwe, B., and Müller, Th. (eds). *Towards a New Epistemology of Mathematics*, special issue of *Erkenntnis*, 68:3 (2008).

Esfeld, M., Ledwig, M., and Spohn, W. (eds). *Current Issues in Causation*, Paderborn: Mentis (2001).

Freitag, W., Rott, H., Sturm, H., and Zinke, A. (eds). *Von Rang und Namen. Philosophical Essays in Honour of Wolfgang Spohn*, Paderborn: Mentis (2016).

Freitag, W. and Zinke, A. "The theory of form logic," in: *Logic and Logical Philosophy*, 21 (2012), pp. 363–389.

Fuhrmann, A. and Olsson, E. (eds). *Pragmatisch denken*, Frankfurt: Ontos (2004).

Haas-Spohn, U. (ed.). *Intentionalität zwischen Subjektivität und Weltbezug*, Paderborn: Mentis (2003).

Halbach, V. (ed). *Methods for Investigating Self-Referential Truth*, special issue of *Studia Logica*, 68:1 (2001).

Halbach, V. and Olsson, E. (eds). *Coherence and Dynamics of Belief*, special issue of *Erkenntnis*, 50 (1999).

Halbach, V. and Horsten, L. (eds). *Principles of Truth*, Frankfurt; Hänsel-Hohenhausen (2002); Frankfurt: Ontos ($^2$2004).

Hinzen, W. and Rott, H. (eds). *Belief and Meaning – Essays at the Interface*, Frankfurt: Hänsel-Hohenhausen (2002).

Horák, V. and Rott, H. (eds). *Possibility and Reality – Metaphysics and Logic*, Frankfurt: Ontos (2003).

Kahle, R. and Schroeder-Heister, P. (eds). *Proof-Theoretic Semantics*, special issue of *Synthese*, 148:3 (2006).

Kahle, R., Stärk R., and Schroeder-Heister, P. (eds). *Proof Theory in Computer Science* (= Lecture Notes in Computer Science; 2183), Berlin: Springer (2001).

Olsson, E. J. (ed.). *The Epistemology of Keith Lehrer*, Dordrecht: Kluwer (2003).

Olsson, E. J. (ed.). *Belief Revision*, special issue of *Studia Logica*, 73:2 (2003).

Olsson, E. (ed.). *Knowledge and Inquiry: Essays on the Pragmatism of Isaac Levi* (= Cambridge Studies in Probability, Induction and Decision Theory), Cambridge: Cambridge UP (2006).

Olsson, E. J., Schroeder-Heister, P., and Spohn, W. (eds). *Logik in der Philosophie*, Heidelberg: Synchron (2005).

Raidl, E. "Completeness for counter-doxa conditionals—using ranking semantics," in: *Review of Symbolic Logic*, 12:4 (2019), pp. 861–891.

Raidl, E. "Open-minded orthodox Bayesianism by epsilon-conditionalisation," in *British Journal for the Philosophy of Science*, 71:1 (2020), pp. 139–176.

Rott, H. *Change, Choice and Inference: A Study of Belief Revision and Nonmonotonic Reasoning* (= Oxford Logic Guides; 42), Oxford: Oxford UP (2001).

Rott, H., and Williams, M.-A. (eds). *Frontiers in Belief Revision* Dordrecht: Kluwer (2001).

Spohn, W. *Causation, Coherence, and Concepts. A Collection of Essays*, Dordrecht: Springer (2009).

Spohn, W. *The Laws of Belief. Ranking Theory and its Philosophical Applications*, Oxford: Oxford UP (2012); received the Lakatos Award 2012.

Urchs, M. "Complementary Explanations," *Synthese*, 120 (1999), pp. 137–149.

Zinke, A. "A BULLET for invariance: Another argument against the invariance criterion for logical terms," in: *The Journal of Philosophy*, 115:7 (2018), pp. 382–388.

Zinke, A. *The Metaphysics of Logic*, Frankfurt: Klostermann (2018).

### Era Horsten: since 2019

On the one hand, it is obviously way too early to engage in a retrospective and summarize Leon Horsten's time at the university. One the other hand, we cannot *not* mention him and his team; so we do, albeit briefly.

Leon Horsten (born 1966) was born in the Netherlands, completed his MA in Minnesota (1989), and obtained his PhD from the KU Leuven (1993), where he also held his first faculty appointment. It was during his time in Leuven that he first came into contact with Konstanz due to his cooperation (continuing to the present day) with Halbach. From Leuven he moved to Bristol (2008–2019) from where he came to Konstanz, fearing that the Brexit might put an end to his productive years in England.

Before Horsten came, however, the department had abandoned his former concentration on the philosophy of science as its primary mission, and the denomination of Horsten's chair was changed to read "professor of theoretic philosophy with special emphasis on metaphysics, epistemology, and logic." What the department did not change was the expectation that topics in the three areas listed should be addressed by an analytic philosopher which made Horsten, who came as a highly accomplished scholar and an internationally recognized expert for bringing formal tools to bear on philosophical problems, a great fit. He had written, among others, a book in metaphysics: *The Metaphysics and Mathematics of Arbitrary Objects* (2019), had edited a book in (the) epistemology (of mathematics): *Gödel's Disjunction: The Scope and Limits of Mathematical Knowledge* (2016, with Philip Welch), and about twenty of his research papers were published in journals of the ASL. Better-known names among his collaborators (beyond Halbach) are Igor Douven, Graham E. Leigh, Hannes Leitgeb, Øystein Linnebo, Richard Pettigrew, and Philip Welch.

Horsten's initial hires (to fill the two faculty lines associated with his chair) were Mount and Roberts.

Beau Mount did his DPhil in the philosophy of mathematics (*The kinds of mathematical objects*, 2017), supervised by Halbach and Timothy Williamson, and was a junior research fellow at New College (2018–2020) before he came to Konstanz.

Sam Roberts completed his PhD requirements with the thesis *Potentialism and reflection principles* (Birkbeck College, 2016, part of Linnebo's project "Plurals, Predicates, and Paradox"), from where he moved to work at ConceptLab (Oslo, 2017–2020), one of whose co-directors is Linnebo.

The total number of articles all three have published in logic journals suggests that their future research will be closer to core areas of mathematical logic than that of any group before. This orientation can also be found in the doctoral colloquium they run jointly with Carolin Antos (see §5.5) since 2020.

**More logic & foundations in philosophy: Rosenthal**

Jacob Rosenthal (born 1969) earned an MA in mathematics (Würzburg, 1995) before he came to Konstanz to work on the propensity interpretation of probability, supervised by Spohn (2002, published as *Wahrscheinlichkeiten als Tendenzen. Eine Untersuchung objektiver Wahrscheinlichkeitsbegriffe*, 2004). He had competing interests in action theory and ethics, so he left for Bonn and completed a habilitation on action theory (2012, sponsored by Andreas Bartels), published 2017 as *Entscheidung, Rationalität und Determinismus*). In 2013 he came back to Konstanz on a visiting position, and while he holds the chair for practical philosophy since 2016, he has never really stopped thinking about probability.

**Selected publications.**

Rosenthal, J. "Probabilities as Ratios of Ranges in Initial-State Spaces," in: *Journal of Logic, Language and Information*, 21:2 (2012), pp. 217–236.

Rosenthal, J. "Johannes von Kries's Range Conception, the Method of Arbitrary Functions, and Related Modern Approaches to Probability," in: *Journal for General Philosophy of Science*, 47:1 (2016), pp. 151–170.

**More logic & foundations in philosophy: Müller**

Thomas Müller (born 1969) earned an MSc in physics, a PhD in philosophy (both Freiburg, 1997 and 2001) and completed his habilitation in Bonn (2008). His dissertation was on tense logic (*Arthur Priors Zeitlogik. Eine problemorientierte Darstellung*, published 2002) and his habilitation thesis was on moral aspects of promising (*Versprechen. Zur Struktur einer moralischen Praxis*). During his time as assistant professor in Bonn he was not, however, affiliated with the the Department of Philosophy itself but with its *Lehr- und Forschungsbereich III*, formerly known as the Hasenjaeger-Institut, the *Seminar für Logik und Grundlagenforschung*, whose director at the time was Rainer Stuhlmann-Laeisz.[9] Müller first moved to Utrecht (2007–2013), before he accepted the offer for the position as "professor of philosophy with a special emphasis on theoretical philosophy." His job description (i. e., theoretical philosophy) is currently all that is left of what was once the hallmark of philosophy in Konstanz, namely, the philosophy of science; it has been stricken from all chair denominations since. (I was told the department has meanwhile decided to add "philosophy of science" to the denomination for a future replacement.)

Before Müller came to Konstanz, he had distinguished himself by contributions to temporal logic and its metaphysics as well as by spearheading a philosophy of mathematical practice. Most of his more recent research expands on the former and not only develops new improved formalisms for branching time but also teases out the consequences they have for loaded issues such as (in)determinism, free will and agency, open future, and the metaphysics of possibility and time in general. While the metaphysics does not seem to fall on DVMLG territory, the formalisms involved surely do.

**Notable projects, conferences, & guests.** Müller brought two projects from Utrecht to Konstanz: "What is really possible? Philosophical explorations in branching-history-based real modality," funded by the NWO, the Dutch Research Council, and its sister project, funded by the ERC, the European Research Council, "Indeterminism Ltd. An intervention on the free will debate." The first project employed Marius Backmann, Antje Rumberg, and Rosja Mastop, while the second saw Michael De, Verena Wagner, Niels van Miltenburg, Antje Rumberg, Jesse Mulder, and Daan Evers as post-docs or graduate students. Those who contributed to the logic were Rumberg and De.

Antje Rumberg, who wrote her MA thesis on Bolzano under direction of Schroeder-Heister, received her PhD from Utrecht University (*Transitions toward a semantics for real possibility*, 2016), supervised by Albert Visser and Müller. She worked with Müller in Konstanz (2013–18) before she moved on, first to Stockholm, then Aarhus.

Michael De, who hails from Canada and did his MA at Simon Fraser, earned his PhD, supervised by Stephen Read and Peter Milne, at St Andrews (*Negation in context*, 2011). He worked with Müller in Utrecht (2011–2013) and Konstanz (2013–2015 & 2016–2017 as an adjunct), before me moved via Miami and Bern to Utrecht.

---

[9]This unit, specifically created for Hasenjaeger in 1962, was terminated when Stuhlmann-Laeisz retired in 2008. Cf. E. Brendel & R. Stuhlmann-Laeisz, *Geschichte des Lehrstuhls für Logik und Grundlagenforschung an der Rheinischen Friedrich-Wilhelms-Universität Bonn*, in this volume.

The last activities that deserve mention are ongoing and not yet completed. There is, first, the DFG-funded project "Different kinds of conditionals: Coin tosses and kangaroos in the forest of alternative possibilities," which is a follow-up to the project already mentioned: "Alternatives for the future" (see page 74). It is executed by Hadil Karawani, a trained linguist, who wrote her PhD thesis on counterfactuals with Frank Veltman (*The real, the fake, and the fake fake*, Amsterdam, 2014). The goal is to achieve a unified analysis "based on a (modal) forest of (temporally tree-like ordered) alternative possibilities." Second, there are the twin projects "Suspending Belief" (*Sich doxastisch enthalten*, 2020–2022), along with the participation in the international effort "Thinking About Suspension" (2021–2024). The latter is coordinated not by Müller but by Wagner (Konstanz) and Zinke (see page 71).

Verena Wagner, who did her doctoral work with Rott (*Free and coerced agency: A new approach to classical compatibilism*, Regensburg, 2013,) cooperated with Müller already on the earlier project "Indeterminism Ltd." She plans to turn her research in the context of the two projects into a habilitation.

In the past, the analysis of counterfactuals and the modeling of doxastic states required some formal machinery at the intersection of formal semantics and logic; whether this applies to the three projects just mentioned is in at least one of their possible futures.

**Selected publications.**
Belnap, N., and Müller, Th. "CIFOL: Case-Intensional First Order Logic (I): Toward a Theory of Sorts," in: *Journal of Philosophical Logic*, 43:2–3 (2014), pp. 393–437.

Belnap, N., and Müller, Th. "BH-CIFOL: Case-Intensional First Order Logic (II): Branching Histories," in: *Journal of Philosophical Logic*, 43:5 (2014), pp. 835–866.

Belnap, N., Müller, Th., and Placek, T. *Branching Space-Times. Theory and Applications*, Oxford: Oxford UP (2022).

De, M. "Intrinsicality and counterpart theory," in: *Synthese*, 193:8 (2016), pp. 2353–2365.

De, M. and Omori, H. "Classical Negation and Expansions of Belnap–Dunn Logic," in: *Studia Logica*, 103:4 (2015), pp. 825–851.

Müller, Th. "Time and Determinism," in: *Journal of Philosophical Logic*, 44:6 (2015), pp. 729–740.

Müller, Th., Rumberg, A., and Wagner, V. (eds). *Real possibilities, indeterminism, and free will: Three contingencies of the debate*, special issue of *Synthese*, 196:1 (2019).

Rumberg, A. "Transition Semantics for Branching Time," in: *Journal of Logic, Language and Information*, 25:1 (2016), pp. 77–108.

## 4 Logic in mathematics
### Era Prestel: 1975–2008

**People: their stories & their projects.** Alexander Prestel (born 1941) did his doctoral work in number theory with Karl-Bernhard Gundlach in Münster (*Die elliptischen Fixpunkte der Hilbertschen Modulgruppen*, 1966) and his habilitation, promoted by Gisbert Hasenjaeger, in Bonn (*Untersuchungen über Pasch-freie Geometrien und semi-geordnete Körper*, 1972), where he and Ronald Jensen were assistant professors and Prestel became the successor to Wolfram Schwabhäuser. The official denomination of his position in Konstanz was algebra, but (according to one contemporary witness) the intent of the search committee was to hire a logician, and they believed Prestel to be an expert in higher set theory. In the spirit of the original idea that mathematics should be part of the university's service-teaching unit (the *Interfakultät*), physicists urged that new hires in mathematics should be for applied mathematics; they were appeased with the argument that mathematical logic is applied mathematics: mathematics applied to logic.

The other permanent member of Prestel's group was Ulf Friedrichsdorf whom he got to know as a student during his time as an assistant professor in Bonn. Friedrichsdorf had left Bonn for Kiel on a doctoral fellowship to work with Klaus Potthoff and became the latter's only doctoral student (*Existentiell abgeschlossene und generische Zahlstrukturen*, 1973). But the position as assistant professor in Kiel he had hoped for was sacked, so when Prestel inquired whether he would like to come to Konstanz, Friedrichsdorf moved south. His hire proved critical for the reason mentioned already in previous sections: having secured himself a permanent lecturer position early on, Friedrichsdorf could devote himself to an academic life that embodied the inter- and transdisciplinary spirit which people on the *Gründungsausschuss* meant to be a hallmark of the university. While Prestel devoted his time and energy exclusively to mathematics, Friedrichsdorf taught or team-taught many interdisciplinary logic classes and was the face of logic to people outside the department. And while all graduate students were strictly speaking students of Prestel (Friedrichsdorf lacked the credentials (habilitation) to officially supervise graduate students), a really nice masters thesis, written by Christopher von Bülow on Rosser sentences (and later published as a book), was written under Friedrichsdorf's supervision and forced none less than Bob Solovay to admit a slip and to correct a proof.

Among the longer-term assistants were Koenigsmann, Schmid, and Schweighofer.

Jochen Koenigsmann came with a BA in mathematics and philosophy from Oxford in order to do his PhD in Konstanz under Prestel's supervision. His research was centered around model theory and arithmetic of fields where he also co-supervised numerous master theses in real algebraic geometry, in valuation theory, and on Hilbert's 10th Problem. A break through during his time at Konstanz (1985–2001) was his Galois characterization of $p$-adically closed fields (1995) which had been an open problem ever since, in 1965, Ax-Kochen and, independently, Ershov had provided the $p$-adic analogue of Tarski's results on the model theory of real closed fields. From Konstanz he moved back to Oxford, where, inspired by his teaching around Hilbert's 10th Problem in Konstanz, he gave a diophantine definition, not for $\mathbb{Z}$ in $\mathbb{Q}$ (which would prove Hilbert's 10th Problem for $\mathbb{Q}$ to be unsolvable), but for the complement of $\mathbb{Z}$ in $\mathbb{Q}$ (see *Annals of Mathematics*, 183:1, 2016, pp. 73ff.). Koenigsmann was an invited speaker at the ICM 2018 in the section on "Logic and Foundations."

Jürgen Schmid (born 1959) completed his PhD in 1991 (*Existentiell abgeschlossene Integritätsbereiche mit reellen Radikalrelationen*), work which resulted in co-authored publications with Prestel. He wrote his habilitation thesis in Dortmund (1999) and obtained the *venia legendi* in Konstanz (1999). While he started to teach as *Gymnasiallehrer* in 2000, he has been teaching as an adjunct professor since 2004. His teaching and research interests are, besides topics in mathematical logic proper, quadratic forms and associative rings.

Markus Schweighofer (born 1974) did his doctoral work on *Iterated rings of bounded elements and generalizations of Schmüdgen's theorem* (2002) and wrote a cumulative habilitation thesis entitled *Positive polynomials, sums of squares and optimization* (2007). Two years later he obtained a tenured position as professor in Konstanz. He and his doctoral students (six former, three current) keep working on various aspects of polynomial optimization and organize conferences and workshops in the area. The most recent cooperation is the European POEMA project (2019–22), and recent work showed, for the first time by applying techniques from logic and real closed fields to a Lasserre's hierarchy, that many systems of real polynomial inequalities can be converted to slightly generalized linear programs.

**Teaching, research, students.** *Teaching.* Prestel and Friedrichsdorf taught the usual slate of introductions to or upper-level classes on logic and set theory and offered courses for graduating seniors and seminars for doctoral students. On a semi-regular basis Prestel taught a cycle of three classes: mathematical logic, model theory, and recursion theory, but left it mostly to Friedrichsdorf to supplement it with a class on set theory which was offered outside that cycle and hence more often. These classes followed the usual pattern of two hours of lecture each week combined with either a two-hour tutorial (*Übungen*) for introductory-level classes or a two-hour seminar for upper-level classes. Prestel made a name for himself as a good and effective lecturer, known for his famous 'two-sponge method' (cleaning the chalk board with two wet sponges simultaneously to shave off more lecture time). And although the logic classes were advertised as a good stepping stone towards a masters thesis (*"zusammen mit anderen Vorlesungen und Seminaren aus meinem Zyklus kann sie auf dem Gebiet der Modelltheorie zu einer Staats- und Diplomarbeit führen"*), most graduate work was done in algebra. Beyond logic, Prestel lectured on algebra (abstract, linear, real), algebraic curves, algebraic geometry, function fields, and number theory. (For Friedrichsdorf, see the remarks in previous sections.) From the mid-1980's to the mid-1990's Prestel und Friedrichsdorf took their students (as Ziegler did from Freiburg) to the famous Läuchli–Specker Seminar in Zürich. (It was started by Bernays and Gonseth in 1936; Läuchli died unepectedly in 1997 but Specker continued until 2002.) Since this required an overnight stay in Zürich, it was as much a social event as it was an academic experience.

*Research.* Prestel's research was in model theory, or more specifically, in the model theory of fields (including developments that stemmed from Hilbert's 10th and 17th problems) to whose toolkit he added more algebraic methods, e.g., valuation theory or infinite Galois theory from the arithmetic of fields. Although he attended the workshop in Oberwolfach on Shelah's work when it first hit (1972; then again in 1980), Prestel did not concern himself with 'pure' model theory such as stability theory or the resulting classifications efforts. To the contrary, his own focus and the research direction his students took gravitated more and more towards algebra, a development that is also reflected in how his research group was named and in his book titles (see below). Initially, and for many years, Prestel called his group simply "Logic and Algebra." Later this was spelled out further by listing real algebraic geometry, Galois theory, valuation theory, and model theory. Eventually, in 2006, it became simple again: "Real Geometry and Algebra." Colleagues report that his lecture notes on formally real fields and formally $p$-adic fields (with Roquette) have triggered quite some research interest in the community and think that the joint paper with Schmid from 1990, which continued earlier work on preordered fields from 1982, is an exceptionally nice piece of work, in which the authors axiomatize and prove the decidability of the theory of the ring of algebraic integers. Prestel seemed to have shared the sentiment: he made it the topic of one out of the two talks he ever gave at Oberwolfach (the other was on positive polynomials in 1997).

*Students.* Among Prestel's 17 doctoral students—Wilfried Meißner (1979), Margarita Delso (1987), Camilla Grob (1988), Joachim Schmid (1991), Maria Pia Solèr (1993), Thomas Jacobi (1999), Maria Eugênia Canto Cabral (2005), Holger Merkel (2009), Sven Wagner (2009), Sabine Burgdorf (2011), and Samuel Volkweis Leite (2013)—at least six continued beyond their doctoral work and obtained faculty positions: Bernhard Heinemann (1982; Hagen, Germany), Cydara Ripoll (1991; Universidade Federal do Rio

Grande do Sul, Brazil), Jochen Koenigsmann (1993; Oxford), Mihai Prunescu (1998; IMAR, Bucharest), Markus Schweighofer (2002; Konstanz), and Tim Netzer (2008; Innsbruck, Austria). If we use the MSC classification as a guideline, then most dissertations fell into abstract algebra (primary classification 12–16, with clusters at 12 (field theory and polynomials) and 13 (commutative algebra)), while three have a traditional logic flair: Grob (decidability in certain closed fields), Meißner (model theory of quadratic forms), and Prunescu (diophantine definability and Matiyasevich's theorem); two more, Cabral and Wagner, worked on effective decision procedures (whether certain quadratic modules are Archimedean). Some of Prestel's students added work in logic or computer science logic (MSC classification 03 and 68, respectively) later in their career: Heinemann (03, 68), Koenigsmann (03), Prunescu (03, 68), and Schmid (03, 68).

**Notable projects, conferences, & guests.** Major research projects with Prestel as PI were "Arithmetic of fields" (internally funded), "Representation of positive polynomials" (2000–2006, DFG-funded), and a project on approximating nonnegative polynomials (*Selbstkonkordante Barrieren für Kegel nichtnegativer Polynome*, 2005–2008); the latter funded research conducted by Markus Schweighofer and later by Tim Netzer. Prestel joined the research training group "Mathematical logic and its application" (2002–2008), co-founded by his former student Jochen Koenigsmann and done in cooperation with nearby Freiburg (its chair was Jörg Flum), whose focus was on finite model theory and its applications for computer science (*Graduiertenkolleg #806*, which is some kind of fixed-term graduate school funded by the DFG). Later Prestel became a member of the European research network "Real Algebraic and Analytic Geometry" (RAAG, 2002–2006), among whose main objectives was "the training of young researchers through active participation in research of the highest quality." This last cooperation had a lasting impact on the direction of his research group.

Prestel quite regularly attended the annual logic meetings at Oberwolfach from 1970 to 1983. But around the time their focus shifted more towards proof theory and constructive mathematics, he started to organize workshops on model theory that met every other year: 1982 (mit W. Baur and A. Macintyre), 1984 (with L. van den Dries and U. Felgner), 1986 (with G. Cherlin), 1988 (with W. Hodges and M. Ziegler), 1990 (with L. van den Dries and P. Roquette), 1992 (with D. Lascar and M. Ziegler), 1994 (with U. Felgner and M. Ziegler), 1998 (with Y. Ershov and M. Ziegler), and 2000 (with D. Lascar and M. Ziegler). In addition he participated in the Oberwolfach meetings on the Arithmetic of Fields (1990, 1993, 2002, 2006, 2009, 2013), Real Algebraic Geometry (1984, 1993, 1997), Quadratic Forms (1978, 1985, 1995), and Valued Fields (2010).

Prestel's co-organizers, the organizers of other workshops he attended, and recurring participants at both indicate the research cooperations he maintained. Members of his network include, among others, Peter Roquette (Heidelberg), Martin Ziegler (Freiburg), Eberhard Becker (Dortmund), Wulf-Dieter Geyer (Erlangen), and Manfred Knebusch (Regensburg) within Germany; outside Germany there were Antonio Engler (Campinas, Brazil), Yuri Ershov (Paris, then Novosibirsk), Moshe Jarden with his students Dan Haran and Ido Efrat (Tel Aviv), Max Dickmann, Françoise Delon and Zoé Chatzidakis (Paris), Angus Macintyre and Alex Wilkie (Oxford), Gregory Cherlin (Rutgers), Lou van den Dries (Utrecht, then Urbana), and Charles Delzell (Stanford, then Louisiana). All visited Konstanz, some for a year or longer (e.g., Delzell, Efrat, Engler, Ershov, Haran, van den Dries).

**Selected publications.**

Berr, R., Delon, F., and Schmid, J. "Ordered fields and the ultrafilter theorem," in: *Fundamenta Mathematica*, 159:3 (1999), pp. 231–241.

Friedrichsdorf, U. *Einführung in die klassische und intensionale Logik*, Braunschweig: Vieweg (1992).

Friedrichsdorf, U., and Prestel, A. *Mengenlehre für den Mathematiker* (= Grundkurs Mathematik), Braunschweig: Vieweg (1985).

Klep, I. and Schweighofer, M. "Connes' embedding conjecture and sums of Hermitian squares.," in: *Advances in Mathematics*, 217:4, pp. 1816–1837; addendum *ibid.* 252 (2014), pp. 805–811.

Kriel, T. L. and Schweighofer, M. "On the exactness of Lasserre relaxations and pure states over real closed fields," in: *Foundations of Computational Mathematics*, 19:6 (2019), pp. 1223–1263.

Koenigsmann, J. "From $p$-rigid elements to valuations (with a Galois characterization of $p$-adic fields)," in: *Journal für die reine und angewandte Mathematik*, 465 (1995), pp. 165–182.

Prestel, A. "Pseudo real closed fields," in: *Set Theory and Model Theory* (= Lecture Notes in Mathematics; 872), ed. R. B. Jensen and A. Prestel, Berlin: Springer (1981), pp. 127–156.

Prestel, A. *Einführung in die Mathematische Logik und Modelltheorie* (= Grundkurs Mathematik), Braunschweig: Vieweg (1986); rev. engl. tr. Delzell, C. N.: *Mathematical Logic and Model Theory. A Brief Introduction* (= Universitext), London: Springer (2011).

Prestel, A. and Delzell, C. N. *Positive Polynomials: From Hilbert's 17th Problem to Real Algebra* (= Springer Monographs in Mathematics), Berlin: Springer (2001).

Prestel, A. and Roquette, P. *Formally p-adic Fields* (= Lecture Notes in Mathematics; 1050), Berlin: Springer (1984).

Prestel, A. and Schmid, J. "Existentially closed domains with radical relations. An axiomatization of the ring of algebraic integers," in: *Journal für die reine and angewandte Mathematik*, 407 (1990), 178–201.

Prestel, A. and Schmid, J. "Decidability of the rings of real algebraic and $p$-adic algebraic integers," in: *Journal für die reine and angewandte Mathematik*, 414 (1991), 141–148.

Prestel, A. and Ziegler, M. "Modeltheoretic methods in the theory of topological fields," in: *Journal für die reine and angewandte Mathematik*, 299/300 (1978), 318–341.

Schmid, J. "Regularly $T$-closed fields," in: *Hilbert's Tenth Problem: Relations With Arithmetic and Algebraic Geometry* (= Contemporary Mathematics; 270), ed. J. Denef et al., Providence: AMS (2000), pp. 187–212.

Schweighofer, M. "Iterated rings of bounded elements and generalizations of Schmüdgen's Positivstellensatz," in: *Journal für die reine and angewandte Mathematik*, 554 (2003), pp. 19–45.

## Era Kuhlmann: since 2009

**People: their stories & their projects.** Salma Kuhlmann (born 1958) completed her college education at McGill University in Montréal, did her PhD work with Daniel Lascar at the Université Paris Diderot–Paris VII (1991, *Quelques propriétés des espaces vectoriels values en théorie des modèles*), and obtained the habilitation with her work on *Ordered Exponential Fields* at the University of Heidelberg in 1999. Before she joined faculty ranks in Konstanz—from a tenured position at the University of Saskatchewan, where she remains to hold an adjunct position—she had visiting positions at Punjab University, Chandigarh (1995), and the Fields Institute, Toronto (1996–1997). She describes her research interests as to falling into five areas: (i) model-theoretic algebra: o-minimal structures, saturated and recursively saturated o-minimal expansions; (ii) model theory of valued fields: ordered fields, fields of power series, dependent fields, exponential fields, Hardy fields, exponential-logarithmic power series fields, transexponential fields; (iii) ordered algebraic structures: lexicographic orderings, ordered vector spaces, Hahn groups; (iv) models of arithmetic: integer parts; and (v) real algebraic geometry: *Positivstellensätze*, moment problems, symmetric positive polynomials.

The first two post-docs who stayed on for some time were Carl and Infusino.

Merlin Carl obtained his PhD with Peter Koepke in Bonn (*Alternative finestructural and computational approaches to constructibility*, 2011), after which he came to Konstanz for six years (2011–2017) before he moved on to Flensburg. He worked, among others, on infinite-time computation, diophantine polynomials encoding proofs, exponential real closed fields, and taught introductory courses to mathematical logic but also branched out into philosophy (Husserl, Gödel, Lakatos). The legacy, however, he created for himself was the interdisciplinary (computer science, law, linguistics, mathematics, and philosophy) logic colloquium "Logic in Konstanz" that he initiated, at the beginning co-sponsored by Spohn. Kuhlmann and Antos (see § 5.5) took over when Carl left.

Maria Infusino hailed from Italy, where she had worked, under supervision of Aljosa Volčič in Calabria, on her dissertation *Uniform distribution of sequences of points and partitions* (2011). She spent three years in Reading (2011–2014) before she joined Kuhlmann's group (2014–2020). Her interests lie at the intersection of analysis and real algebraic geometry as well as their application to mathematical physics. She was awarded her habilitation degree in 2020 and returned to Calabria for a tenure-track position shortly thereafter.

Currently, three more post-docs seek additional academic qualification, be it their habilitation or otherwise; two are 'home-grown', Krapp and Serre, while Brickhill came from Bristol.

Hazel Brickhill did her PhD, advised by Philip Welch, in set theory (*Generalising the notions of closed unbounded and stationary sets*, 2017), for which she received a Faculty of Science Commendation, and continues research along those lines; in a side job she has teamed up with Horsten (§ 3), whom she knows from her time as a graduate student, to explore a theory of non-standard infinitesimal probabilities via ultrafilter constructions.

Lothar Sebastian Krapp, who came to Konstanz after completing his MMath with Jonathan Pila in Oxford, wrote his dissertation on *Algebraic and model-theoretic properties of o-minimal exponential fields* (2019), did joint work with Carl on exponential fields, and intends to further investigate ordered algebraic structures for his habilitation.[10]

Michele Serra came with a combined masters degree in mathematics from Leiden and Padova (by courtesy of the Erasmus Mundus Joint Masters program) and completed his doctoral work on *Automorphism groups of Hahn groups and Hahn fields* in 2021. His research interests include, more broadly, commutative algebra, valuation theory, and ordered structures.

In addition to these five assistant professors in her group, she won two fellows for the *Zukunftskolleg*: Margaret Thomas (2010–2019) and Pantelis Eleftheriou (2013–2021); cf. §§ 5.3 & 5.4.

---

[10] He was selected as one of the speakers at the PhD Colloquium at *Colloquium Logicum 2022* in Konstanz. The PhD Colloquium is the DVMLG's celebration of the most talented PhD students in logic: the members can nominate candidates for the PhD Colloquium talks among those who graduated after the last *Colloquium Logicum* and before the next. The Programme Committee of the *Colloquium Logicum* then selects the speakers. An invitation to speak at the PhD Colloquium can be seen as the German "best dissertation in logic" award.

A few junior people did not stay for long.

Annalisa Conversano (2009–2011) had written her doctoral thesis in Siena, supervised by Alessandro Berarducci in Pisa: *On the connections between definable groups in o-minimal structures and real Lie groups: the non-compact case* (2009), and left after two years to assume duties as Senior Lecturer in Auckland. While in Konstanz, her closest cooperation was with Anand Pillay.

Itay Kaplan (2010–2011) is a student of Saharon Shelah (*Topics in dependent theories*, 2009) and returned to Jerusalem after one year in Konstanz.

Mickaël Matusinski, a frequent co-author of Kuhlmann, passed through for a brief stay (2009–2010). He had done his doctoral work in Dijon with Jean-Philippe Rolin, *Ordinary differential equations with coefficients in a field of generalized power series* (2007), and had worked with Kuhlmann before in Saskatchewan (2008). He is now in Bordeaux.

**Teaching, research, students.** *Teaching.* Kuhlmann's team members continue the Prestel-Friedrichsdorf tradition and offer the full gamut of introductory logic classes (set theory, model theory, and recursion theory) but, also as before, treat proof theory as the red-haired step child. Clearly, they continued to offer the algebra courses (linear and abstract) mandatory for majors. What Kuhlmann seems to do differently is that she offers lecture classes that teach knowledge prerequisite for a meaningful participation in research seminars (e.g., introduction to real algebraic geometry) on a more regular basis and supported by her own lecture notes. Members of her team have likewise enriched the existing curriculum. Infusino, for instance, quite regularly taught graduate classes on topological vector spaces and topological algebras along with one-offs such as positive polynomials and moment problems.

*Research.* In the preceding section we briefly indicated the research interests that members of Kuhlmann's group pursue and note that as the research group became more international and more mixed, so did the range of topics being studied. As we saw above, towards the end of his career, Prestel considered himself an algebraist who had delegated logic to Friedrichsdorf. This trend away from logic was reversed by Kuhlmann. While Prestel's *Oberseminar reelle Geometrie und Algebra* was maintained—the German *Oberseminar* is a weekly meeting for graduate and post-graduate students to discuss new literature and their own ongoing research—within a year it saw a companion-*Oberseminar Modelltheorie* (2010) to include Frehm (see below) and Thomas, which, as more new members joined the team, first morphed into *Mathematische Logik, Mengenlehre und Modelltheorie* (2018) when Brickhill joined, and then into "Complexity Theory, Model Theory, Set Theory" (2020) after Mateusz Michałek joined the department as professor (chair) of real algebraic geometry. Contemporary research in mathematical logic, it seems, is represented more broadly these days.

Michałek joined the group, but being a chair himself, he is independent of it. The same was true for Fehm. – Arno Fehm completed his PhD in Tel Aviv in 2010 (*Decidability of large fields of algebraic numbers*) under supervision of Moshe Jarden (as part of the European Research and Training Network "Galois Theory and Explicit Methods"). While working on his doctoral thesis, he was in frequent contact with Ziegler, Koenigsmann, and Prestel. So he joined the faculty in Konstanz as a *Juniorprofessor* (2010–2016), after which he went to Manchester for a year before he returned to TU Dresden on a tenured position. He did mostly his own thing while in Konstanz, but participated in Kuhlmann's *Oberseminar* and organized, with Pierre Dèbes (Lille) and Lior Bary-Soroker (Tel Aviv) the French-German summer school "Galois Theory and Number Theory." With Prestel, who at the time still attended the *Oberseminar* as an emeritus, he co-authored a joint paper; it was received August 29, 2014, and is the last paper Prestel submitted.

*Students.* Since Kuhlmann came in 2009, a number of students completed their PhD under her supervision. Besides Krapp and Serra, mentioned above, the other doctoral students were Charu Goel: *Extension of Hilbert's 1888 theorem to even symmetric forms* (2014); Katharina Dupont: *Definable valuations on NIP fields* (2015); Simon Müller: *Quasi-ordered rings: A uniform study of orderings and valuations* (2020). The dissertation by Gabriel Lehéricy: *Quasi-ordres, C-groupes, et rang différentiel d'un corps différentiel valué* (2018), was co-directed with Françoise Point (Sorbonne), and co-directed with Infusino was Patrick Michalski: *A systematic approach to infinite-dimensional moment problems* (2012). The MSC classification 03, it seems, is back prominently.

When it comes to fostering student success, we should mention that Infusino, as a member of the association "European Women in Mathematics" and with Kuhlmann's support, founded a local chapter "Konstanz Women in Mathematics" (KWIM) and ran events from 2013 through 2019. When Infusino left, she passed the baton to Kuhlmann.

*Conferences.* Kuhlmann is no stranger to the MFO (*Mathematisches Forschungszentrum Oberwolfach*—she has been invited to meetings on model theory since 1998 and organized meetings at the MFO herself (2014, 2017, 2020)—but it is just one venue in addition to events at the university, which are open to everyone. She, members of her team (e.g., Antos, Eleftheriou, Goel, Infusino, Thomas), and external partners: Paolo D'Aquino (Naples), Alessandro Berarducci (Pisa), Philip Ehrlich (Ohio), Didier Henrion (Toulouse), Tobias Kuna (Reading), Jonathan Pila (Oxford), and Victor Vinnikov (Ben Gurion), have been organizing about two meetings a year, many not at Konstanz. Kuhlmann and Thomas (see §5.3) have founded the annual regional (Basel, Freiburg, Konstanz, Passau) workshop on model theory, called *Donau-Rhein Modelltheorie und Anwendungen*, which has met every year since its inauguration in 2017.

**Selected publications.**

Carl, M. "Optimal results on recognizability for infinite time register machines," in: *Journal of Symbolic Logic*, 80:4 (2015), pp. 1116–1130.

Carl, M. and Krapp, L. S. "Models of true arithmetic are integer parts of models of real exponentiation," in: *Journal of Logic & Analysis*, 13:3 (2021) pp. 1–21.

Chernikov, A. and Kaplan, I. "Forking and dividing in $NTP_2$ theories," in: *Journal of Symbolic Logic*, 77:1 (2012), pp. 1–20.

Conversano, A. and Pillay, A. "Connected components of definable groups and o-minimality I," *Advances in Mathematics*, 231:2 (2012), pp. 605–623.

D'Aquino, P., Knight, J., Kuhlmann, S., and Lange, K. "Real closed exponential fields," in: *Fundamenta Mathematicae*, 219 (2012), pp. 163–190.

Dupont, K., Hasson, A., and Kuhlmann, S. "Definable valuations induced by multiplicative subgroups and NIP fields," in: *Archive for Mathematical Logic*, 58:7–8 (2019), pp. 819–839.

Fehm, A. and Prestel, A. "Uniform definability of Henselian valuation rings in the Macintyre language," in: *Bulletin of the London Mathematical Society*, 47:4 (2015), pp. 693–703.

Ghasemi, M., Infusino, M., Kuhlmann, S., and Marshall, M. "Moment problem for symmetric algebras of locally convex spaces," in: *Integral Equations and Operator Theory*, 90:3 (2018), art. 29 (19 pp.).

Goel, C., Kuhlmann, S., and Reznick, B. "The analogue of Hilbert's 1888 theorem for even symmetric forms," in: *Journal of Pure and Applied Algebra*, 221:6 (2017), pp. 1438–1448.

Infusino, M. "Quasi-analyticity and determinacy of the full moment problem from finite to infinite dimensions," in: *Stochastic and Infinite Dimensional Analysis* (= Trends in Mathematics), ed. C. C. Bernido et al., Basel: Birkhäuser (2016), pp. 161–194 (= ch. 9).

Infusino, M. and Kuhlmann, S. "Infinite dimensional moment problem: Open questions and applications," in: *Ordered Algebraic Structures and Related Topics* (= Contemporary Mathematics; 697), ed. F. Broglia et al., Providence: AMS (2017), pp. 187–201.

Krapp, L. S., Kuhlmann, S., and Serra, M. "On Rayner structures," in: *Communications in Algebra*, 50:3 (2022), pp. 940–948.

Kuhlmann, S. and Matusinski, M. "Hardy-type derivations in generalized series fields," in: *Journal of Algebra*, 351 (2012), pp. 185–203.

Lehéricy, G. "On the structure of groups endowed with a compatible C-relation," in: *Journal of Symbolic Logic*, 83:3 (2018), pp. 939–966.

## 5 Zukunftskolleg

The *Zukunftskolleg* started operation in 2001, initially as a three-year pilot scheme called "Center for Junior Research Fellows" (the ZWN, *Forschungszentrum für den wissenschaftlichen Nachwuchs*), and was instituted in its current form in 2007 as an Institute for Advanced Study for early-career researchers. Its mission has remained the same, namely, to counteract the incrusted structures at German universities that can leave junior faculty at a considerable disadvantage. The basic idea is straightforward: if a post-doc can find the grant money, the university will provide the support structure. Since 2007, the first time Konstanz was recognized as one of the Top 10 in Germany (Excellence Initiative), the *Zukunftskolleg* disposes of enough federal grant money to award its own fellowships.

### 5.1 Luc Bovens & probability (2002–2005)

Among the first to take advantage of the *Zukunftskolleg* was Bovens. He returned to Konstanz on a Sofja Kovalevskaja Award (DFG) to form a research group, co-directed with Hartmann, so that they could continue their cooperation on all things Bayesian. The group was named "Philosophy, Probability, and Modeling" (2002–2005), and their range of topics was staggering: from "Bayesian networks in philosophy" to "Models of terrorism prevention." Topics were arranged into four groups: (i) Evidence and Confirmation, (ii) Rational and Social Choice, (iii) Probabilistic Causation, and (iv) Uncertain Reasoning. Project leads were—they were all post-docs—Claus Beisbart, Franz Dietrich, Armond Duwell, Ludwig Fahrbach, Natalie Gold, Amit Hagar, Franz Huber, Luca Moretti, Veiko Palge, Gabriella Pigozzi, Robert Bishop, Rolf Haenni, Iain Martel, Christoph Schmidt-Petri, and Paul Thorn; almost all of them obtained permanent faculty positions later. Group members organized 20 conferences, workshops, and summer schools, 13 of them met in Konstanz, and produced about 30 working papers. Prominent visitors over the summer included James Hawthorne, Christian List, Miklós Rédei, and Teddy Seidenfeld.

**Selected publications.**

Beisbart, C., Bovens, L., and Hartmann, S. "A utilitarian assessment of alternative decision rules in the council of ministers," in: *European Union Politics*, 6:4 (2005), pp. 395–418.

Bovens, L. and Hartmann, S. *Bayesian Epistemology*, Oxford: Oxford UP (2003).

Dietrich, F. "How to reach legitimate decisions if the procedure is controversial," in: *Social Choice and Welfare*, 24 (2005), pp. 363–393.

Dietrich, F. and List, Ch. "The impossibility of unbiased judgment aggregation," in: *Theory and Decision*, 68 (2010), 281–299.

Dietrich, F. and Moretti, L. "On coherent sets and the transmission of confirmation," in: *Philosophy of Science*, 72:3 (2005), pp. 403–424.

Fahrbach, L. and Hartmann, S. "Normativität und Bayesianismus," in: *Deskriptive oder normative Wissenschaftstheorie*, ed. B. Gesang, Frankfurt: Ontos (2005), pp. 177–204.

Gold, N. and List, Ch. "Framing as path dependence," in: *Economics & Philosophy*, 20:2 (2004), pp. 253–277.

Haenni, R. and Hartmann, S. "Modeling partially reliable information sources: A general approach based on Dempster–Shafer theory," in: *Information Fusion*, 7 (2006), pp. 361–379.

Hartmann, S. and Pigozzi, G. "Judgment Aggregation and the Problem of Truth-Tracking," in: *Proceedings of the 11th Conference on Theoretical Aspects of Rationality and Knowledge (TARK XI)*, New York: ACM (2007), pp. 248–252.

Huber, F. "Assessing theories, Bayes style," in: *Synthese*, 161:1 (2008), pp. 89–118.

Huber, F. "The logic of theory assessment," in: *Journal of Philosophical Logic*, 36:5 (2007), pp. 511–538.

## 5.2 Franz Huber & formal epistemology (2008–2013)

Franz Huber (born 1977) did his MA in logic with Paul Weingartner and Johannes Czermak (Salzburg, 2000) and his PhD with Gerhard Schurz in Erfurt (2004, *Assessing theories. The problem of a quantitative theory of confirmation*). He was a member of the Bovens–Hartmann group at Konstanz (2002–2005), worked at CalTech (2005–2007), before he returned to Konstanz to head the research group "Formal Epistemology" (2008–2013). It was located at the *Zukunftskolleg* and funded by an Emmy Noether fellowship (DFG). And while it may look as though Huber simply continued the work done earlier in Konstanz by Fuhrmann, Olsson, Rott, and Spohn, this was not the case, at least not initially, when Schurz and Bovens had a greater influence. It was only over time that Huber moved closer to Spohn's ideas.

Members of the group worked on the following projects: (i) Knowledge and justification (Peter Brössel, PhD *Rethinking Bayesian Confirmation Theory*, 2012); (ii) Belief and its revision (Benjamin Bewersdorf; PhD, same title, 2012); (iii) Degrees of belief and belief (Zinke,[11] Huber); (iv) Theories of degrees of belief (Huber); (v) Degrees of rational acceptability (Anna-Maria Eder, PhD *A study on the foundations of theories of epistemic rationality*, 2016); (vi) Belief revision in dynamic epistemic logic and ranking theory (Peter Fritz; he left after his BA to continue in Amsterdam, then Oxford); (vii) Understanding normality (Corina Strößner). As indicated by the theses that were completed, the research group did double-duty as a graduate program affiliated with the department of philosophy. Strößner was the only post-doc (PhD *Logic and semantics of normality statements*, supervised by Niko Strobach, Saarbrücken, 2012), while Katharina Felka (not linked to any one project), obtained her MA and then went to Hamburg for her PhD.

The group hosted "Monthly Monday Meetings," often with guest speakers from abroad, and (co-)organized six international conferences in formal epistemology (four of them lovingly called 'Formal Epistemology Festival'); external co-organizers included Ray Briggs, Igor Douven, Kenny Easwaran, Branden Fitelson, Eric Swanson, and Jonathan Weisberg.

**Selected publications.**
Brössel, P. and Eder, A.-M. "How to resolve doxastic disagreement," in: *Synthese* 191:11 (2014), pp. 2359–2381.

Brössel, P., Eder, A.-M., and Huber, F. "Evidential support and instrumental rationality," in: *Philosophy and Phenomenological Research*, 87:2 (2013), pp. 279–300.

Huber, F. "Structural equations and beyond," in: *Review of Symbolic Logic*, 6:4 (2013), pp. 709–732.

Huber, F. "New foundations for counterfactuals," in: *Synthese*, 191:10 (2014), 2167–2193.

Huber, F. "What should I believe about what would have been the case?," in: *Journal of Philosophical Logic*, 44:1 (2015), pp. 81–110.

Huber, F., Swanson, E., and Weisberg, J. (eds). *Conditionals*, special issue of *Erkenntnis*, 70:2 (2009).

## 5.3 Margaret Thomas & o-minimal structures (2010–2019)

Margaret Thomas (in Konstanz: 2010–2017) obtained her PhD, supervised by Alex Wilkie in Oxford, with a thesis on *Convergence and parameterization in o-minimal structures* (2009). She stayed in Konstanz for nine years, before she landed a tenure-track position first at McMaster (2018), then at Purdue (2020). During her time in Kostanz she supported herself primarily through a DFG grant "Parameterization and Algebraic Points in O-Minimal Structures" (2012–2017). While in Konstanz, beyond polishing her docto-

---

[11] See page 71; this was her MA thesis.

ral work (jointly with Wilkie), she extended and branched out from it; a frequent research collaborator was Gareth O. Jones, once a fellow graduate student in Oxford. She did some teaching and participated in Kuhlmann's *Oberseminar*, but otherwise cooperation was limited to the organization of workshops: with Kuhlmann she organized the annual *Donau-Rhein Modelltheorie und Anwendungen*, with Eleftheriou and Kuhlmann the "Summer School in Tame Geometry" (2016), and with Antos "European Women in Mathematics" (Graz, 2018). Thomas was the sole organizer of the international workshop "O-Minimality and Applications" (2015). She advised one PhD thesis: Derya Çıray: *Mild parameterization in o-minimal structures* (2019).

**Selected publications.**
Andújar Guerrero, P., Thomas, M., and Walsberg, E. "Directed sets and topological spaces definable in o-minimal structures," in: *Journal of the London Mathematical Society* (2), 104 (2021), pp. 989–1010.

Chernikov, A., Starchenko, S., and Thomas, M. "Ramsey growth in some NIP structures," in: *Journal of the Institute of Mathematics of Jussieu*, 20:1 (2021), pp. 1–29.

Jones, G. O. and Thomas, M. "Effective Pila-Wilkie bounds for surfaces implicitly defined from Pfaffian functions," in: *Mathematische Annalen*, 381 (2021), pp. 729–767.

Jones, G. O., Thomas, M., and Wilkie, A. "Integer-valued definable functions," in: *Bulletin of the London Mathematical Society*, 44 (2012), pp. 1285–1291.

## 5.4 Pantelis Eleftheriou & recovering structures (2013–2021)

Pantelis Eleftheriou got his PhD from Notre Dame under the supervision of Sergei Starchenko (*Groups definable in linear o-minimal structures*, 2007) and came to Konstanz in 2013 after post-docs in Barcelona (2007–2008), Lisbon (2008–2011), and Waterloo (2011–2013); first on a two-year fellowship, followed by a five-year one, both at the *Zukunftskolleg*. He obtained the post-doctoral degree of habilitation with the thesis *Structure theorems and applications in semi-bounded and tame pairs* in 2019. In 2021 he moved to Leeds for a tenure-track position. His research interest is "to recover concrete mathematical structure from given logical data, such as Lie groups from definable groups, algebraic curves from definable sets with many rational points, algebraically closed fields from strongly minimal structures, and algebraic topology (homotopy/homology) from semi-linear data." While in Konstanz, he cooperated repeatedly with Ayhan Günaydin (Istanbul), Assaf Hasson (Ben Gurion), Philipp Hieronymi (Urbana), and Ya'acov Peterzil (Haifa); he supervised one doctoral student (Alex Savatovsky, *Structure theorems for d-minimal expansions of the real additive ordered group and some consequences*, 2020) and mentored two visiting post-docs: Eliana Barriga and Erick García Ramírez. He participated in Kuhlmann's *Oberseminar* and assisted with nine conferences, two of them local for which he secured the funding: "Summer School in Tame Geometry" (2016) and "Tame Expansions of o-Minimal Structures" (2018).

**Selected publications.**

Eleftheriou, P. "Semi-linear stars are contractible," in: *Fundamenta Mathematicae*, 241 (2018), pp. 291–312.

Eleftheriou, P., Günaydin, A., and Hieronymi, P. "Structure theorems in tame expansions of o-minimal structures by a dense set," in: *Israel Journal of Mathematics*, 239 (2020), pp. 435–500.

Eleftheriou, P., Hasson, A., and Keren, G. "On definable Skolem functions in weakly o-minimal non-valuational structures," in: *Journal of Symbolic Logic*, 82 (2017), pp. 1482–1495.

Eleftheriou, P., Hasson, A., and Peterzil, Y. "Strongly minimal groups in o-minimal structures," in: *Journal of the European Mathematical Society*, 23 (2021), pp. 3351–3418.

### 5.5 Carolin Antos & set theory (2016–2023)

Carolin Antos (in Konstanz: from 2016) did her MA and PhD in Vienna at the Kurt Gödel Research Center for Mathematical Logic (formerly, the *Institut für Logistik*) with Sy Friedman (*Foundations of Higher-Order Forcing*, 2016). In 2016 she came to the *Zukunftskolleg* on two successive grants and has her own research group since 2018 "Forcing: Conceptual Change in the Foundations of Mathematics" (2018–2023), generously founded by a Freigeist Fellowship of the Volkswagen Foundation. The overarching goal of the project on forcing is twofold: first, to track how the technique of forcing has transformed set theory as a discipline, and second, to analyze the ramifications this has for mathematics, its foundations and its philosophy. Members of her group are Neil Barton, Deborah Kant, and Daniel Kuby. Barton did doctoral work, advised by Ian Rumfitt, on set theory at Birkbeck College (*Executing Gödel's programme in set theory*, 2016), from where he went to Vienna (2016), to Konstanz (2019), to Oslo (2022). He and Antos know each other from their time as graduate students (he used her results at critical junctures in his dissertation) and seem to form the hub of the group. Kant, well-known among those who have an interest in the philosophy of mathematical practice, works on questions of independence and naturalness in set theory, while Kuby, who did doctoral work on Paul Feyerabend (*Studies on Paul Feyerabend's philosophy: from logical empiricism to the historical turn in philosophy of science*, 2017) supervised by Elisabeth Nemeth in Vienna, pursues a satellite project in which he frames the universe-multiverse debate in set theory as a philosophy-of-science question about intertheoretic inconsistency. Associated members are Regula Krapf (Bonn), who did her doctoral work on class forcing (Bonn, 2017),[12] and Nick de Hoog (graduate student, co-supervised with Hamburg). Thus far, group members have organized four major conferences.

But Antos is doing more than directing a research group. She became an assistant professor (with special emphasis on the philosophy of mathematics) at the Department of Philosophy. As such, she teaches introductory logic classes mandatory for philosophy majors but also team-teaches graduate classes with members of the Kuhlmann group. Moreover, after Carl had left, she joined in the organization of the interdepartmental logic group "Logic in Konstanz," which hosts semi-regular meetings with internal and external speakers. She and her activities align perfectly with what the *Gründungsausschuss* once envisioned for all of Konstanz: truly interdisciplinary research.

**Selected publications.**

Antos, C., Barton, N., and Friedman, S. "Universism and extensions of V," in: *The Review of Symbolic Logic*, 14:1 (2021), pp. 112–154.

Antos, C., Barton, N., Friedman, S., Ternullo, C., and Wigglesworth, J. (eds). *Foundations of Mathematics*, special issue of *Synthese*, 197:2 (2020).

Barton, N. "Forcing and the universe of sets: Must we lose insight?" in: *Journal of Philosophical Logic*, 49:4 (2020), pp. 575–612.

Barton, N., Müller, M., and Prunescu, M. "On representations of intended structures in foundational theories," in: *Journal of Philosophical Logic*, online first (September, 2021).

Barton, N., Ternullo, C., and Venturi, G. "On forms of justification in set theory," in: *The Australasian Journal of Logic*, 17:4 (2020), pp. 158–200.

Centrone, St., Kant, D., and Sarikaya, D. (eds). *Reflections on the Foundations of Mathematics. Univalent Foundations, Set Theory and General Thoughts* (= Synthese Library; 407), Cham: Springer (2019).

Kant, D. and Sarikaya, D. "Mathematizing as a virtuous practice: Different narratives and their consequences for mathematics education and society," in: *Synthese*, 199:1–2 (2021), pp. 3405–3429.

---

[12] Krapf was a speaker at the PhD Colloquium at the *Colloquium Logicum 2018* in Bayreuth; cf. Footnote 10.

# Zum Zermelo-Ring

## Heinz-Dieter Ebbinghaus[1], Benedikt Löwe[2,3,4]

[1] Mathematisches Institut, Albert-Ludwigs-Universität Freiburg, Ernst-Zermelo-Straße 1, 79104 Freiburg im Breisgau, Deutschland

[2] Institute for Logic, Language and Computation, Universiteit van Amsterdam, Postbus 94242, 1090 GE Amsterdam, Niederlande

[3] Fachbereich Mathematik, Universität Hamburg, Bundesstraße 55, 20146 Hamburg, Deutschland

[4] Churchill College, Lucy Cavendish College, & Department of Pure Mathematics and Mathematical Statistics, University of Cambridge, Storey's Way, Cambridge CB3 0DS, England

E-Mail: hde@uni-freiburg.de, loewe@math.uni-hamburg.de

**Vorbemerkung.** Dieser Artikel besteht aus drei Teilen: im ersten Teil berichtet der erste Autor (Heinz-Dieter Ebbinghaus) über den Hintergrund des Rings, den Ernst Zermelo von seinem Großvater geerbt hat, und wie die Idee eines Preises entstanden ist; im zweiten Teil berichtet der zweite Autor (Benedikt Löwe) darüber, wie die DVMLG den Zermelo-Ring als Preis ausgestaltet hat; der dritte Teil ist der offizielle durch den Vorstand der DVMLG am 19. Juli 2022 beschlossene Text der Preisauslobung.

**Der Ring** (Heinz-Dieter Ebbinghaus).

Gegen Ende der 1990er Jahre, bei einem meiner Besuche, schenkte mir Zermelos Witwe Gertrud (1902–2003) einen Ring, den ihr Mann oft getragen habe. Es ist der Siegelring seines Großvaters Ferdinand Zermelo, daher trägt er die Initialen "FZ" auf der Siegelfläche (vgl. Abb. 1).

Ferdinand Zermelo, ein gelernter Buchbinder, übernahm 1839, fünf Jahre nach der Geburt von Zermelos Vater Theodor, eine Buch-, Kunst- und Schreibwarenhandlung in Tilsit. Er beteiligte sich intensiv am kulturellen und sozialen Leben der Stadt. So gehörte er zu den Gründern der Tilsiter Schiller-Gesellschaft und des Tilsiter Kunstvereins. Mehrfach hatte er das Amt eines Stadtrats inne.[1] Seine Liebe zur klassischen Literatur findet sich bei Zermelos Vater und bei Zermelo wieder. Gertrud Zermelo berichtete, ihr Mann habe ihr an manchen Abenden aus literarischen Werken vorgelesen. Sie deutete an, dass Zermelos Wertschätzung des Siegelrings die Wertschätzung spiegelt, die er seinem Großvater entgegengebracht hat.

Getragen habe ich den Ring nur einmal, meiner Erinnerung nach bei einem Vortrag, den ich über Zermelo gehalten habe. Lange war mir nicht

---

[1] Vgl. H.-D. Ebbinghaus, *Ernst Zermelo. An Approach to His Life and Work.* Springer-Verlag 2007, Kapitel 1.

ABBILDUNG 1. Der Siegelring Ferdinand Zermelos. (Bild: Heinz-Dieter Ebbinghaus.)

klar, was mit ihm geschehen solle. Angeregt durch den Iffland-Ring,[2] kam mir im Sommer 2002 der Gedanke, der Ring könne einem wissenschaftlichen Preis den Namen geben. Er könne dadurch an das wissenschaftliche Werk Zermelos erinnern, aber auch daran, wie intensiv Gertrud Zermelo ihrem Mann in den letzten Jahren seines Lebens zur Seite gestanden hat, in einer Zeit, die durch seine Erblindung und durch Enttäuschung über die seiner Meinung nach ausbleibende Anerkennung seiner wissenschaftlichen Arbeit geprägt war.

Gertrud Seekamp (so ihr Name vor der Heirat) und Zermelo lernten sich im März 1934 kennen, als Zermelo in den sogenannten Bernshof nach Freiburg-Günterstal umzog. Gertrud Seekamp wohnte auch dort, und die beiden haben, wohl von Beginn an, Zeit miteinander verbracht (vgl. Abb. 2 *links*). So hat Ernst Zermelo ihr Interesse an klassischer Literatur geweckt. Im Jahre 1944 heirateten die beiden; danach kümmerte sich Gertrud Zermelo intensiv um ihren Gatten, der mit zunehmender Blindheit immer mehr auf ihre Hilfe angewiesen war. Darüber hinaus hat sie sich auch anderer Belange angenommen. So hat sie, als die Schweiz in den ersten Jahren nach dem Krieg keine Pensionszahlungen an Zermelo überwies, mit Hilfe deutscher Behörden erreicht, dass wenigstens ein Teil der Gelder zur Ausgabe in der Schweiz bereitgestellt wurde. Das ermöglichte dem Ehepaar eine gemeinsame Reise in das Tessin (vgl. Abb. 2 *rechts*). Ferner hat sie den Umzug der beiden noch lebenden Schwestern Zermelos, Elisabeth (verstorben 1952) und Margarete Zermelo (verstorben 1959), aus Südtirol in ein Freiburger Altenheim ermöglicht.

---

[2]Der *Iffland-Ring*, ein diamantbesetzter Eisenring mit dem Porträt des Schauspielers August Wilhelm Iffland (1759–1814) wird von seinem Träger testamentarisch an den seiner Meinung nach bedeutendsten und würdigsten männlichen Bühnenkünstler des deutschsprachigen Theaters auf Lebenszeit verliehen. Der Ring ist Eigentum der Republik Österreich, und sein derzeitiger Träger ist Jens Harzer (geboren 1972).

ABBILDUNG 2. *Links*. Gertrud Seekamp (ab 1944 Zermelo) und Ernst Zermelo im August 1935 im Park des Bernshofes (Villa Dr. Berns) in Freiburg/Breisgau. *Rechts*. Ernst und Gertrud Zermelo im Sommer 1948 im Tessin. (Bilder: Zermelo-Nachlass, Universitätsarchiv Freiburg.)

Zum Tod Zermelos am 21. Mai 1953 schreibt Paul Bernays an Gertrud Zermelo gleichsam eine Zusammenfassung:

> Sie, liebe Frau Zermelo, können das befriedigende Bewusstsein haben, dass Sie Ihrem Mann in seinen späten Lebensjahren viele Mühen des Daseins abgenommen und ihn vor Verlassenheit und trauriger Einsamkeit bewahrt haben. Dafür werden auch alle, die der markanten und originellen Forscherpersönlichkeit Ihres Mannes würdigend gedenken, Ihnen Dank wissen.

Am 3. September 2002, ihrem hundertsten Geburtstag, habe ich Frau Zermelo von meinen Vorstellungen über den Ring berichtet. Sie war sehr erfreut. In der Folge habe ich öfter darüber nachgedacht, wie diese Vorstellungen einer Realisierung näher zu bringen seien. Es war die Zeit, in der die körperlichen Kräfte Frau Zermelos bei gleichbleibender geistiger Frische deutlich abnahmen, eine Entwicklung, die Ende 2003 zu ihrem Tod führte.

Frei von dem zeitlichen Druck, Frau Zermelo eine Realisierung des Preises noch erleben zu lassen, entschied ich mich, Überlegungen über seine Ausgestaltung bis zur Fertigstellung der Biografie Zermelos und der Herausgabe seiner gesammelten Werke zurückzustellen. Als das geschehen war, wandte ich mich an die DVMLG und schlug schlicht einen nach dem Zermelo-Ring benannten wissenschaftlichen Preis vor, dessen Ausgestaltung bei Annahme dieses Vorschlags in den Händen der DVMLG liegen sollte. Mein stiller

ABBILDUNG 3. Gertrud Zermelo im Jahre 1999 im Hause der Familie Ebbinghaus. (Bild: Heinz-Dieter Ebbinghaus.)

Wunsch war es, dass bei der Vergabe des Preises auch die Bedeutung Gertrud Zermelos für die Altersjahre Zermelos sichtbar werden könnte.

Den Ring habe ich inzwischen in die Obhut des Universitätsarchivs Freiburg gegeben, wo sich der übrige Nachlass Zermelos befindet. Es ist ausdrücklich vorgesehen, dass die DVMLG bei der Gestaltung des Preises auf ihn zurückgreifen kann.

**Der Preis** (Benedikt Löwe).

Die Idee der Ausgestaltung eines mit dem Siegelring Ferdinand Zermelos verbundenen Preises wurde im September 2012 an die DVMLG herangetragen.[3] Es hat insgesamt zehn Jahre gedauert, bis diese Idee in die Wirklichkeit umgesetzt wurde. In diesen Jahren wurde das Thema dreimal auf Mitgliederversammlungen eingehend und z. T. kontrovers diskutiert.[4]

Die Mitglieder der DVMLG hatten zwei grundsätzliche Sorgen bei der Ausgestaltung des Preises:

1. Eine tatsächliche Übergabe des Rings wie beim Iffland-Ring kam nicht in Frage, da nicht gewährleistet hätte werden können, daß der Ring für die nächste Übergabe wieder zur Verfügung steht; es wurde also

---

[3] E-mail von Heinz-Dieter Ebbinghaus an Ulrich Kohlenbach v. 15. September 2012. Es gab bereits einen früheren Vorschlag: "eine ... Angelegenheit ..., die ich bereits vor einigen Jahren an die DVMLG herangetragen habe, die dann aber nach anfänglichen Überlegungen nicht weiter verfolgt worden ist".

[4] Protokoll der Mitgliederversammlung in Neubiberg v. 4. September 2014; Protokoll der Mitgliederversammlung in Bayreuth v. 15. September 2018; Protokoll der elektronischen Mitgliederversammlung v. 3. März 2021.

über die Anfertigung von Repliken und deren Kosten diskutiert. Die Kosten der Anfertigung einer Replik alle vier Jahre wurde von vielen Mitgliedern als unverhältnismäßig im Vergleich zum Jahreshaushalt der DVMLG angesehen.

2. Man wollte vermeiden, daß der Preis in Konkurrenz mit bereits existierenden Preise aus dem Bereich der Grundlagen der Mathematik steht (z.B. mit dem *Sacks Prize* „awarded annually for the best doctoral dissertation in mathematical logic" oder dem *Karp Prize* „awarded for an outstanding paper or book in the field of symbolic logic most of which has been completed in the time since the previous prize was awarded").

Nach eingehender Diskussion beauftragte die Mitgliederversammlung am 3. März 2021 den Vorstand mit der Entscheidung über die Details der Preisausgestaltung. Der Vorstand der DVMLG beschloß am 19. Juli 2022 die Auslobung des Preises alle vier Jahre; die durch den Vorstandsbeschluß festgelegte Beschreibung des Preises findet sich unten im Text.

**Offizielle Beschreibung des Preises.** Die üblicherweise verwendete axiomatische Grundlage der Mathematik sind die sogenannten Zermelo-Fraenkel-Axiome der Mengenlehre, u.a. benannt nach Ernst Zermelo (1871–1953). Zermelos Werk war von unschätzbarem Wert für die Grundlagen der Mathematik, ging aber weit darüber hinaus und umfaßte außerdem Arbeiten zur Variationsrechnung, angewandten Mathematik und Physik. In den 1990er Jahren überreichte Ernst Zermelos Witwe Gertrud den Siegelring Ernst Zermelos, seinerseits von Ernst Zermelos Großvater Ferdinand geerbt, an Professor Heinz-Dieter Ebbinghaus, der anregte, im Zusammenhang mit diesem Ring einen Preis auszuloben.

**(1)** Die DVMLG verleiht alle vier Jahre den ***Ernst-Zermelo-Ring*** an eine aktive Forscherin oder einen aktiven Forscher in den Grundlagen der Mathematik, die oder der einen nachhaltigen Einfluß auf die Entwicklung des Gebiets gehabt hat und bei der oder dem vorherzusehen ist, daß sie oder er auch in den kommenden Jahren aktiv und einflußreich bleiben wird.

**(2)** Die auszuzeichnende Person wird von einem Preiskomitee bestimmt, welches aus Vertreterinnen und Vertretern des Vorstandes der DVMLG und aus international führenden Forscherinnen und Forschern besteht. Forscherinnen und Forscher aus allen Bereichen der Grundlagen der Mathematik kommen als Preisträgerinnen oder Preisträger in Frage. Eine Beziehung zum Werk Ernst Zermelos ist keine Voraussetzung für die Preisverleihung, kann aber vom Preiskomitee bei der Auswahl berücksichtigt werden.

**(3)** Die Verleihung des Ernst-Zermelo-Rings findet im festlichen Rahmen statt, wenn möglich vor einem breiten Publikum, und beinhaltet eine

Laudatio und einen Festvortrag. Die DVMLG übernimmt die Reise- und Unterkunftkosten der geehrten Person und der anderen unmittelbar an der Verleihungzeremonie beteiligten Personen. Nach der Verleihung und bis zur erneuten Verleihung nach vier Jahren ist die geehrte Person die **Trägerin** oder der **Träger des Ernst-Zermelo-Rings**.

(4) Der Ring wird im Universitätsarchiv Freiburg verwahrt und von dort für die Verleihungzeremonie entliehen. Die Trägerin oder der Träger des Ernst-Zermelo-Rings kann während der vier Jahre bis zur erneuten Verleihung den Ring für besondere Anlässe entleihen.

# DLMPS—Tarski's vision and ours

## Wilfrid Hodges[*]

Herons Brook, Sticklepath, Okehampton EX20 2PY, England

**Historical note (2022)**
This is a slightly updated re-print of the published version of the author's Presidential Address held on 19 July 2011 at the Fourteenth *International Congress of Logic, Methodology and Philosophy of Science* at Nancy, originally published as [13]. It is re-printed in this volume with the kind permission of the author and the publisher. Both DLMPS and IUHPS changed their names at the General Assembly in Helsinki on 6 August 2015 to *Division for Logic, Methodology and Philosophy of Science and Technology* (DLMPST) and *International Union of History and Philosophy of Science and Technology* (IUHPST), respectively (cf. § 4). Since this paper was originally written before this change of name and acronym, the old acronyms DLMPS and IUHPS are used.

**Preliminary remark (2015)**
The title is the title I gave for my talk. Naming individuals enriches history, and Tarski is a natural person to name, both because of his very articulate views about the reasons for doing logic, and also because of his broad and lasting personal influence. In [2, Chapter 10], Solomon and Anita Burdman Feferman give a very readable account of Tarski's role in the setting up of DLMPS. But there is a danger that by naming Tarski I diminished the contributions of many other people whose interests combined to shape DLMPS; I hope the paper itself will set the balance straight.

## 1  What happened fifty years ago

DLMPS, or to give it its full title, the *Division of Logic, Methodology and Philosophy of Science*, held its first international congress in 1960 at Stanford University, California. Starting with the Third International Congress at Amsterdam in 1967, these congresses have taken place every four years. So the 2011 congress is the nearest thing we have to a celebration of the first half-century of DLMPS congresses.

---

[*]For help of various sorts I thank Anne Fagot-Largeault, Efthymios Nicolaidis, Thomas Piecha, Peter Schroeder-Heister, Paul van Ulsen, Henk Visser, Jan Woleński, and the DLMPS Executive Committee of 2008–11. But none of them should be held responsible for views expressed below.

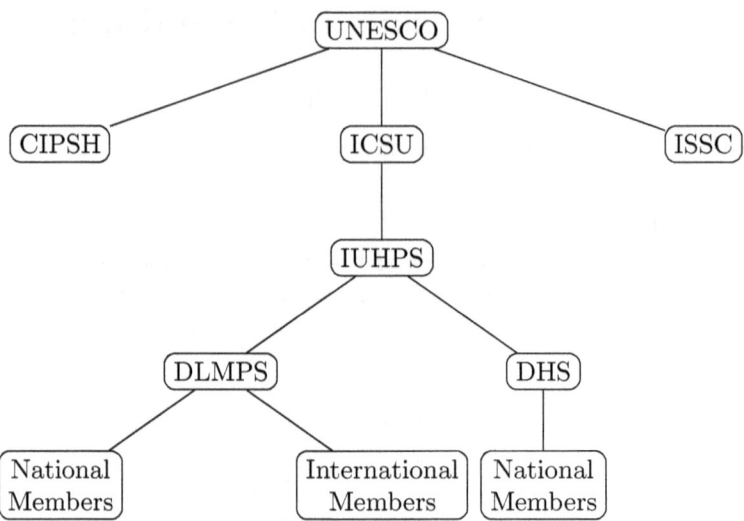

FIGURE 1. The DLMPS in the ICSU family in 1955. For explanations of the acronyms, cf. §1.2. ICSU and ISSC merged in 2018 to form the *International Science Council* (ISC). DHS (now DHST) became a member of CIPSH in 2001; DLMPS became a member of CIPSH in 2011.

The editors of the proceedings of the 1960 Stanford congress (Ernest Nagel, Patrick Suppes and Alfred Tarski) wrote in their preface [16, p. vi]:

> This was the first International Congress for Logic, Methodology and Philosophy of Science since the International Union of the History of Science and the International Union of the Philosophy of Science established the International Union of the History and Philosophy of Science on June 3, 1955. The congresses of a related character held prior to the formation of IUHPS were mainly devoted to the philosophy of science. The title of the 1960 Congress reflects its broader coverage; it was in fact the first international congress to include a large number of papers on both mathematical logic and the methodology and philosophy of science.

The editors refer to the establishment of IUHPS, the *International Union of the History and Philosophy of Science*. In fact, DLMPS came into existence as one of the two Divisions of IUHPS, creating a splatter of acronyms as in Figure 1. Let me run through this figure.

## 1.1 Upwards from ICSU

At the top is UNESCO, the *United Nations Educational, Scientific and Cultural Organization*, which was born in 1946. During the Second World War there had been discussions between countries on the Allied side with a view to setting up supranational organisations after the war. The creation of the United Nations in 1945 was one result of these discussions. Another was UNESCO, which was attached to the United Nations and thus became funded by and answerable to the national governments ratifying the United Nations Charter. The original plan was for UNESCO to support just education and culture; Joseph Needham and Julian Huxley successfully argued that science should be included too [10, p. 71f].

ICSU, the *International Council of Scientific Unions*, had existed since 1931 as an international alliance of scientific organisations.[1] It had grown out of collaborations between the scientific academies of some European countries, together with some international scientific projects such as global distance measurements or the establishment of standards. Because of these mixed origins it had two kinds of member: "national adhering organisations" like the Royal Society, and international scientific unions like the *International Union of Pure and Applied Chemistry*. The aims of ICSU in 1931 were (in summary):

(1) to coordinate member organisations,

(2) to direct other international scientific activity,

(3) to promote science in countries through their national academies.

At the outset the members of ICSU were forty national members and eight international unions [10, Chapter 3].

In 1946, UNESCO and ICSU formally recognised each other. This meant in practice that UNESCO could call on ICSU for scientific expertise, and ICSU could call on UNESCO for money for the kinds of venture likely to appeal to the United Nations. These arrangements still stand; e.g., Rio+20, the 2012 United Nations Conference on Sustainable Development held in Rio de Janeiro, had a strong input from UNESCO and ICSU together.

## 1.2 Downwards from ICSU

The next step down from ICSU in the diagram is IUHPS, the *International Union of History and Philosophy of Science*. There had been an *International Academy of the History of Science* as early as 1928. When UNESCO came into being, Needham and others felt that an *International Union of*

---

[1] In 1998, ICSU changed its name to *International Council for Science* while retaining the acronym ICSU; in 2018, ICSU merged with the *International Social Science Council* (ISSC) to form the *International Science Council* (ISC).

*the History of Science* would be a valuable addition to ICSU. So UNESCO negotiated with the International Academy to convert it into the IUHS, which duly became a member of ICSU in 1947.[2]

In 1946, responding to a suggestion of Józef Bocheński who pointed to the recently-formed *Association for Symbolic Logic* and its associated *Journal of Symbolic Logic*, Ferdinand Gonseth (a Swiss mathematician with interests in philosophy of science and the foundations of mathematics) launched the *International Society of Logic and Philosophy of Sciences* (*Société Internationale de Logique et de Philosophie des Sciences*; SILPS) with an associated journal *Dialectica*. His chief colleagues in this were Paul Bernays, Karl Popper, and Karl Dürr. At about the same time, Stanislas Dockx (a Belgian philosopher of science) set up an *International Academy of Philosophy of Science*. When Gonseth and Dockx became aware that the *International Academy of the History of Science* had been converted into a member of ICSU, they decided to pool their efforts so as to create an *International Union of the Philosophy of Science* (IUPS), which would apply to ICSU for membership. So they called a meeting of interested parties in Brussels in July 1949, where plans were made to set up the IUPS. Besides representatives of UNESCO and ICSU, and Robert Feys representing the *Association for Symbolic Logic*, the meeting included the logicians Evert Beth and L. E. J. Brouwer together with several leading European philosophers of science. The inaugural meeting of IUPS took place in Paris in October 1949. Sometime between July and September 1949, presumably under pressure from ICSU which wanted to avoid a proliferation of smaller unions, it was agreed that IUHS and IUPS should amalgamate into a single union. In September the executive of IUHS appointed three delegates, and in October IUPS responded with its own three delegates (Gonseth, Dockx and Raymond Bayer), to meet in Paris in 1950 to draw up statutes for a combined IUHPS. In fact it took until 3 June 1955—the date quoted above—to agree the form of IUHPS, and the new union was admitted to ICSU in August 1955.[3]

The previous paragraph is based on the detailed first-hand account by Dockx [1]. Dockx was writing in honour of Gonseth, and he chose not to mention one embarrassing event. In 1952, there was a coup in IUPS; Gonseth, Dockx and Bayer were all removed from the executive committee, and presumably from the committee to negotiate with IUHS. The new executive consisted of Albert Châtelet, Arend Heyting, Hans Reichenbach, Bocheński and two participants in the July 1949 meeting: Feys and Jean-Louis Destouches. Feys in correspondence gave two reasons for the coup:

---

[2]Cf. [11]; however, several statements about "the Union" in this article are in fact true only of DHS(T), e.g., the list of officers and the list of commissions.

[3]Cf. also [21].

Gonseth's group wanted to steer UNESCO funds to their own pet projects, and "they were interested in rather literary forms of 'Philosophy of Science'". Given the commitments made by Gonseth and Dockx in 1949, neither of these two points are likely to have had much direct impact on the negotiations with IUHS. But we know that the *Association for Symbolic Logic* was unwilling to throw its weight behind the new union until after the coup, so that the coup may have removed a logjam in the negotiations. There was also a perception on the philosophy side that Petre Sergescu, Executive Secretary of IUHS from 1947 till his death in 1954, was against having a combined union.[4]

According to the formula agreed in 1955, IUHS became the *Division of History of Science* (DHS), IUPS became the *Division of Logic, Methodology and Philosophy of Science* (DLMPS), and the two divisions together formed the *International Union of the History and Philosophy of Science* (IUHPS), which became a member of ICSU replacing IUHS.

During the Presidential Address in Nancy, I said that both Divisions seemed to have lost their copies of the IUHPS statutes by the late 1990s if not earlier—which rather nullified the six years that it had taken to draw up the statutes in the first place. I had reported that Lehto had cited from them in his 1998 book [14, p. 75] and had expressed hope that they could be found somewhere. And, indeed, in May 2013, Benedikt Löwe discovered a copy of the statutes, written in French and dated 1962, in an old box containing documents of the German National Committee of the DLMPS. By that time, the two Divisions had agreed on a *Memorandum of Agreement* that described their collaboration in IUHPS; the 1962 statutes were updated accordingly in 2017 and can be found on the IUHPST website. The 1962 statutes describe the aims of the IUHPS as follows:

(1) *établir des rapprochements entre les historiens et philosophes des sciences et entre les institutions, sociétés, revues, etc. consacrées à ces disciplines ou à des disciplines connexes;*

(2) *rassembler les documents utiles au développement de l'Histoire des Sciences et de la Logique, la Méthodologie et la Philosophie des Sciences;*

(3) *prendre toutes les mesures qu'on croira nécessaires ou utiles pour le développement, la diffusion et l'organisation des études et recherches dans les domaines de l'Histoire des Sciences, de la Philosophie des Sciences et des disciplines connexes;*

(4) *organiser les Congrès Internationaux d'Histoire des Sciences et les Congrès Internationaux de Philosophie des Sciences, ainsi que des Colloques Internationaux;*

---

[4]The quotes from Feys are cited after [21].

(5) *contribuer au maintien de l'unité de la science en général et à l'établissement de liens entre les différentes branches du savoir humain;*

(6) *s'efforcer de favoriser le rapprochement entre historiens, philosophes, savants, soucieux des problèmes de méthode et de fondement que posent leurs disciplines respectives.*

This is similar to the aims stated in the DLMPS statutes.[5]

We should briefly bring Figure 1 up to date. In 1987, DLMPS changed the name "National Members" to "Ordinary Members" because of some political sensitivities. At its General Assembly in Beijing in 2005, DHS added "and Technology" at the end of its name and became DHST. And finally in 2011, DLMPS joined CIPSH, the *Conseil International de la Philosophie et des Sciences Humaines*, which in turn is affiliated to UNESCO (our sister division DHST had joined CIPSH in 2001). In some loose sense, CIPSH is to the Humanities as ICSU is to the Sciences. The *International Social Science Council* (ISSC) was the third such organisation, covering the social sciences. ISSC and ICSU merged in 2018 to form the *International Science Council* (ISC).

## 2 Pennies from heaven

The institutional structures by themselves don't give many clues about the motivations driving the whole machine. The motivations that chiefly concern us here are money and scientific research. Again it will be helpful to begin the discussion with diagrams (Figures 2 & 3). The financial situation today is very different from what it was fifty years ago, so we need diagrams to illustrate both the old situation and the new. These diagrams should be read only as broad indications; one can too easily alter the numbers by adjusting the classifications.

We start with the funds that come to DLMPS from ICSU. UNESCO, which gets its money from countries in the United Nations, makes regular subventions to ICSU. The United States of America, although they withdrew from funding UNESCO in 1984 and resumed in 2003, continued to make substantial contributions to the ICSU grant fund separately through its National Science Foundation. The United States of America withdrew funding from UNESCO again in 2012, and it remains to be seen how this affects the funding of ICSU (and CIPSH, which is in a similar position to ICSU).

For several decades, ICSU passed on a large part of these subventions as grants to its member unions without close scrutiny. But in 1996, an external

---

[5]Cf. *Statutes of the Division of Logic, Methodology and Philosophy of Science and Technology of the International Union of History and Philosophy of Science and Technology*, 6 August 2015, published on the website of the DLMPST.

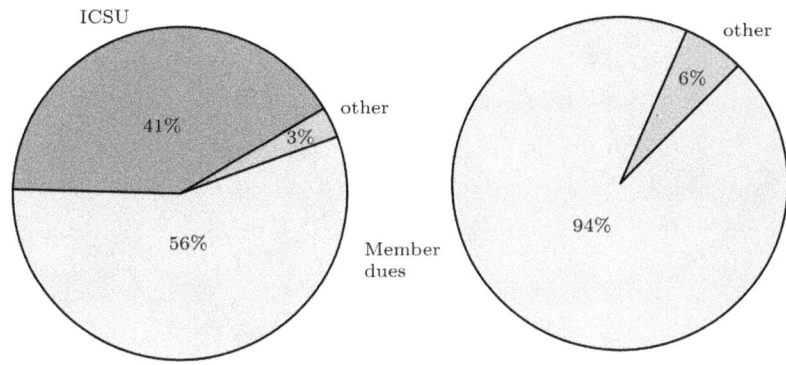

FIGURE 2. DLMPS income. *Left.* 1960s. *Right.* Today (2011).

assessment recommended that ICSU should be more strategic in its allocations.[6] As a result, since 2002 ICSU has awarded grants by competition and peer review, and only for international multidisciplinary ventures in certain announced priority areas. These changes had a dramatic effect on the funding of Unions, as Figure 2 shows for DLMPS. In fact the only grant from ICSU that came to IUHPS since 2002 and before 2014 was a sum in 2004 to allow DHST to set up databases of bibliographical and archival sources. Figure 3 shows the effect on our outgoings. For a while DLMPS supported only its own meetings and some joint activities with DHST, though since 2012 it has also distributed some small grants to conferences sponsored by members. The money that DLMPS puts into the international congresses held every four years is a small fraction of the cost of these congresses, but it serves to prime the pump. In past decades the sale of Congress Proceedings has brought in some income, but today we no longer expect to make any profit on publications.

As supplementary information, it can be reported that one of the eight ICSU grants for 2014 was awarded to IUHPS/DLMPS for a project on *Cultures of Mathematical Research Training.* The grant application was supported by the *International Mathematical Union* and its *International Commission for Mathematical Instruction.* To quote from the project specification:

> This project aims to mobilize the energies of a currently very active research area (the study of *Practice and Cultures of Mathematics*)

---

[6]Cf. *Review of the ICSU Grants Programme, 2001–2006. Report of a CSPR Review Committee,* February 2007. Available on the website of the ISC.

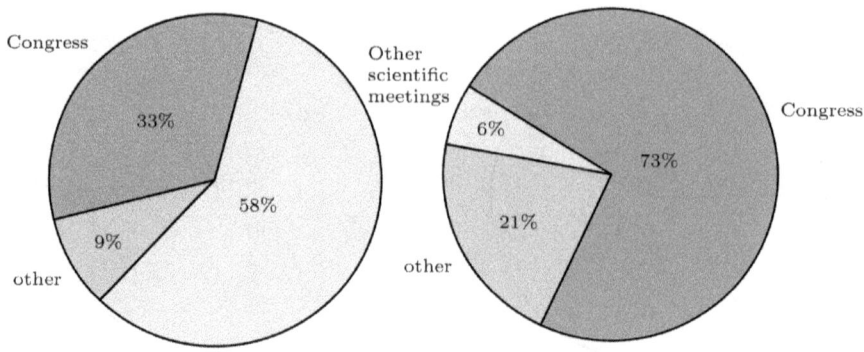

FIGURE 3. DLMPS expenditure. *Left.* 1960s. *Right.* Today (2011).

to provide the theoretical and empirical resources for designing improvements to the training of the next generations of mathematical researchers and the improvement of research education in developing countries.

As Figure 2 shows, virtually all of our present income comes from our members, both Ordinary and International. Common prudence dictates that we should aim to know what these members reckon they are paying for.

## 3  Our members and what they pay for

ICSU has no individual members. In its early days it had only two kinds of member: national bodies and international unions. That was partly because ICSU was, so to say, a meta-level association. Its job was to deal with governments or national academies, and to set up and support scientific associations like the international scientific unions. The unions themselves were not meta-level associations in this sense, but they still tended to have structures that copied those of ICSU. The members of a union would be national committees (often administered either by national scientific academies or by national subject societies) and international scientific societies. Our own union IUHPS is a cipher, but its two divisions still both have this style of membership.

The fact that our members represent societies and institutions means that there is a kind of inertia built into our income: institutions that paid this year are likely to carry on paying next year too, because otherwise they would have to make a decision to stop. This could be dangerous for us, because it tends to hide the question whether we are delivering what our members are paying us for. In fact the position is quite complicated and the remarks below are partly guesswork.

## 3.1 National academies and research councils

About half our members, and two-thirds of our Ordinary Members, are committees of national academies or national research councils. These bodies pass on money from their national governments. Probably most of them reckon that by supporting DLMPS they are supporting science and contributing to the aims of the United Nations as expressed in ICSU. The Canadian National Research Council knows that it is supporting international congresses of DLMPS, and it requests reports from Canadian scientists who attend these congresses; but my impression is that this amount of diligence is very unusual. Some grant-giving bodies ask DLMPS for a copy of our financial report but apparently pay no particular attention to the involvement of logicians or philosophers of science in their countries.

Of course ICSU has its own activities, e.g., government-level conferences like Rio+20. Let me mention two others that are likely to appeal to national academies. The first is the sharing of expertise between different international scientific bodies. Three recent examples are:

(1) In 2010, IUHPS was invited to nominate a member for the advisory board of the annual Gruber Cosmology Prize, worth half a million dollars. We nominated a historian of cosmology proposed by a member of DLMPS Council.

(2) In 2011, IUHPS was invited to support the application of the International Council for Industrial and Applied Mathematics to become a Scientific Associate of ICSU. We sent a positive answer, citing the methodological importance of mathematical modelling.

(3) In 2011, ICSU consulted its members for their comments on its draft ICSU Strategic Plan, 2012–2017. Since the Strategic Plan is largely about environmental issues and the integration of science into governmental planning, IUHPS found nothing to say about it. But perhaps DLMPS should have commented on the proposed *Principle of Universality* for science.

ICSU consultations can be tedious to handle, and often DLMPS is unlikely to have anything to offer. But we could (if membership lists are kept up to date) pass down some consultations to our member societies and national committees. This could help to keep them in touch with the activities of ICSU that they are supporting with their fees.

The second activity of ICSU is its work to protect the free movement of scientists. There is a permanent need for this work, but it was particularly valuable in the days of the Iron Curtain. E.g., DLMPS consulted ICSU for help in getting visas for East European invitees to the Salzburg Congress in 1983.

Besides these activities, ICSU has committees that rely on the unions for their membership. From 2011 to 2014, Maria Carla Galavotti sat on the ICSU Executive Committee; she was nominated by IUHPS on the proposal of DLMPS. In 2005, Deborah Mayo from DLMPS was one of the authors of the ICSU working group report on science and society.[7] In 2008, Susan Lederer became a member of the ICSU Publication Ethics Committee on the proposal of DHST.

## 3.2 Subject societies

There remain the other half of our members, who are not supported by government-funded bodies. Nearly all of these are supported instead by societies devoted to logic or philosophy of science, or both; some are national and some international. It often seems that random factors have decided whether the societies are primarily devoted to logic or to philosophy of science, and it is possible that we have missed out on support in some countries where the logicians and the philosophers of science were not close to each other. We also have only minimal contact with societies of logicians or philosophers of science in South America. The reasons for this are no doubt partly historical, but we observe that our fellow Division has done much better than us in South America; their Congress took place in Rio de Janeiro in 2017.[8]

Our supporting societies represent working logicians and philosophers, and they are more likely to support activities that are directly helpful to these working researchers. In the days when ICSU provided grants, these grants often supported smaller meetings and workshops of the kind that researchers relish. Those days are over, and that's a threat to our income. We saw this, e.g., in Britain in the early 1990s, when the government-funded Royal Society and British Academy stopped paying dues for international unions, and the national committees for these unions had to call on scientific societies instead. The British Logic Colloquium at that date was unable to meet its share of the cost, and for a while Britain dropped to a lower category of membership in DLMPS.

The fact that the international scientific unions don't have individual members comes into play here, because it means that there are no DLMPS scientific activities that individual researchers can feel they are involved in. In fact until 2011, DLMPS was an extreme case. There were just two ways in which individuals could be involved with DLMPS. The first was

---

[7]Cf. *Science and Society: Rights and Responsibilities*. *ICSU Strategic Review*, 2005, available on the website of the ISC.

[8]CLMPST 2023 will be held in Buenos Aires in July 2023.

as officers or members of committees, and the second was as participants in congresses or other meetings organised by DLMPS. The officers had a heavy commitment to DLMPS, and the congress organisers an even greater one, but none of the others did. Participants in meetings registered for the meetings and didn't even need to know what DLMPS is. There were the national committees, but in too many cases the committee had lapsed—we found one case where the committee consisted of one person who had died ten years earlier. Sometimes the only task of these committees was to decide who would be delegates at the four-yearly General Assemblies.

Many of the unions have taken steps to involve individuals in actual scientific work. E.g., the *International Union of Radio Science* (URSI) has ten special-subject commissions and a larger number of working groups. The brief of its *Commission on Radio Astronomy* includes "observation and interpretation of cosmic radio emissions from the early universe to the present epoch".[9] The *International Union of Pure and Applied Physics* has twenty special-subject Commissions; the *Commission on Physics Education* goes back to 1960. I think none of these have an open membership, but they do involve quite large numbers of individuals in more than just bureaucracy. Our fellow Division, DHST, has for many years had special-interest commissions; at least some of them have membership open to any interested individuals, and newsletters are circulated to all members. The DHST website currently (as of 2022) lists five inter-union commissions, thirteen historical commissions, three inter-division commissions, and three scientific sections.

The 2011 General Assembly of DLMPS made a bid to increase the involvement of individual logicians and philosophers of science. It adjusted the statutes so as to allow commissions in the same style as DHST. It set up four commissions, three of them with open membership. One of those three was the Teaching Commission, which has for many years been a commission of DHST and is now an inter-division commission. The other two were new: a *Commission on Philosophy of Technology and Engineering Sciences* and a *Commission on Arabic Logic*. Another inter-division commission followed in 2015, the commission for *History and Philosophy of Computing* (HaPoC). The aim is for DLMPS to make itself more responsive to the needs of researchers.

## 4 The name of the Division

When our sister division added "and Technology" at the end of its name and became DHST, this was a natural step for them to take. The *International Committee for the History of Technology* had been a Scientific Section of DHS since 1968, and several commissions of DHS already had a strong technology component—e.g., the *Scientific Instruments Commission*. So the addition did no more than reflect the facts on the ground.

---

[9]Cf. the website of URSI.

In June 2008, Claude Debru, on behalf of the French National Committee of History and Philosophy of Science, wrote to DLMPS urging us to go down the same road and add "technology" to our scope. We put this to the General Assembly in Nancy in 2011, and the result was a pair of resolutions:

First, the General Assembly agreed in principle that "philosophy of science" in the stated scope of the Division should be expanded to "philosophy of science and technology", and that the Executive Committee should bring to the 2015 General Assembly proposals for changes in the statutes and the name of the Division to give effect to this expansion.

Secondly, the General Assembly asked the Executive Committee to consult with the officers of DHST with a view to changing the name of the Union so as to include technology.

The main reason for proceeding this way was to avoid getting the issue of principle mixed up with debates about the future name of DLMPS. In fact it seemed to many people that just adding T at the end would give a rather monstrous acronym: DLMPST. We tried this acronym on some spell checkers and got back among other things DEMIST, PLUMPEST, ALMOST, DIMMEST and DUMPSITE. Should one or more of the letters be dropped?

## 4.1 Where did L, M, PS come from?

We know what the organisers of the 1960 Stanford Congress thought these letters stood for [16, p. vi]:

> [Stanford] was in fact the first international congress to include a large number of papers on both mathematical logic and the methodology and philosophy of science.

So for the Stanford team, L was for 'mathematical Logic', M was for 'Methodology of science', and PS was for 'Philosophy of Science'.

The name 'Logic, Methodology and Philosophy of Science' could have come from Gonseth back in 1949. Of course if there is evidence against this, then I defer to it; but I know none.

As to mathematical logic: we saw that already in 1947 Gonseth's society was called the International Society of Logic and the Philosophy of Science. Logic was an old interest of Gonseth's. In 1937, he had published a long essay "Qu'est-ce que la logique?" [9, pp. 11-94]. True, that essay was historical rather than mathematical, and even the chapter on Whitehead and Russell's *Principia Mathematica* hardly contains any formulas. But his essay "Philosophie Mathématique" [9, pp. 95–189], published in 1950, is undoubtedly about mathematical logic, including axiomatic set theory and Gödel's incompleteness theorem—even though it does tend to confirm Feys's epithet "rather literary". We might add that although some mathematical logicians were certainly repelled by Gonseth's approach to the subject, others found it a stimulus; Gerhard Heinzmann documents this in the case of Bernays [12].

As to methodology of science: this phrase goes back to the nineteenth century. In Britain it was popularised by William Hamilton of Edinburgh in his lectures in the 1830s and 1840s [15, Appendix; p. 496]:

> The Science of Science, or the Methodology of Science—falls into two branches. ... The former—that which treats of those conditions of knowledge which lie in the nature of thought itself—is Logic, properly so called; the latter,—that which treats of those conditions of knowledge which lie in the nature, not of thought itself, but of that which we think about, ... has been called *Heuretic* ... The one owes its systematic development principally to Aristotle, the other to Bacon.

Speaking in Nancy, it's appropriate to mention that Henri Poincaré used the phrase in the Introduction to his *Science et Méthode* in 1908 [17]:

> *Je réunis ici diverses études qui se rapportent plus ou moins directement à des questions de méthodologie scientifique.*

By the 1940s the notion of scientific methodology was in free circulation among philosophers of science. So it's no surprise that we can document it from Gonseth: *"Essai sur la Méthode Axiomatique"* [4], *"une méthodologie dialectique ouverte"* [5], *"la méthodologie juste en psychologie"* [6], *"La méthodologie des sciences peut-elle être élevée au rang de discipline scientifique?"* [7], and *"Essai sur la Méthodologie de la Recherche"* [8].

In short, the full name "Logic, methodology and philosophy of science" and the parsing of it in the preface to the 1960 Stanford Proceedings could quite easily have come from Gonseth. This is not to say that they would have meant the same to Gonseth as they did to other members of the Division.

## 4.2 A name for the future of DLMPS?

The 2011 General Assembly left it to the new Executive to decide on the future name of the Division. It may be superfluous for me to say anything about it here, but I'll make a few remarks anyway.

The two divisions sit together as representing philosophy of science and technology on the one hand and history of science and technology on the other. So there is no conceivable case for dropping the PS. The situation is different for both the L and the M, but for different reasons.

In the case of M, there is a case for dropping it straight away. The case is that it no longer represents anything distinctive about DLMPS. In the mid 20th century it was common to distinguish methodology from traditional philosophical areas like epistemology and ontology. By advocating "methodology of science", one would be supporting philosophy of science but distancing oneself from metaphysics. E.g., Herbert Feigl published in 1954 a paper with the title "Scientific method without metaphysical presuppositions" [3]. His opening words were:

> As the title of this article indicates, I contend that there are no philosophical postulates of science, i.e., that the scientific method can be explicated and justified without metaphysical presuppositions about the order or structure of nature.

On this interpretation the only reason for retaining the M would be to bracket off certain aspects of the philosophy of science that some people don't want to be associated with. That doesn't strike me as an adequate reason.

Feigl's usage of "method" or "methodology" was not the only one. Tarski had a distinctive view of the matter. His fullest account of it is in the Introduction to the 1941 English version of his book *Introduction to Logic and to the Methodology of Deductive Sciences* [19], and it appears unaltered at least up to the 1961 edition, though it has been shortened in the posthumous 1994 edition.

Tarski distinguishes between "methodology of deductive sciences" and "methodology of empirical sciences". Methodology of deductive sciences is what Tarski elsewhere calls metamathematics (e.g., [18, p. 342]). It is a part of logic, and a part that Tarski strongly associates himself with. Methodology of empirical sciences "constitutes an important domain of scientific research", and logic is valuable for it. But: "logical concepts and methods have not, up to the present, found any specific or fertile applications in this domain" [19, p. xiii]. Tarski comments that this could be a permanent and necessary feature of the subject. He continues:

> It should be added that, in striking opposition to the high development of the empirical sciences themselves, the methodology of these sciences can hardly boast of comparably definite achievements—despite the great efforts that have been made. Even the preliminary task of clarifying the concepts involved in this domain has not yet been carried out in a satisfactory way. Consequently, a course in the methodology of empirical sciences must have a quite different character from one in logic and must be largely confined to evaluations and criticisms of tentative gropings and unsuccessful efforts. [19, p. xiv]

Tarski doesn't spell out what he regards as the tasks of the methodology of empirical sciences—indeed he suggests that some concepts need to be clarified before we can do that properly. But the comparison with metamathematics sends a strong message. A methodologist of an empirical science should ideally aim to find a suitable formal language in which to carry out the science, with suitable meanings for the primitive terms. Then she should look for suitable axioms. Here part of her task will be to find appropriate criteria for the suitability of the axioms. As Tarski explains in [20, p. 366],

> one of the main problems of the methodology of empirical science consists in establishing conditions under which an empirical theory or hypothesis should be regarded as acceptable.

He offers his truth definition as a help here, which suggests that he has in mind a methodologist using a formal metatheory. The oral remarks of Tarski in 1953 reported in [2, p. 250f] point in the same direction.

Tarski makes a few further remarks about "the methodology of empirical science" in [20], but I don't think they help us much here. What is helpful, and perhaps unexpected, is [20, § 19] in which he vigorously dissociates himself from attacks on "metaphysical elements".

> When listening to discussions in this subject, sometimes one gets the impression that the term "metaphysical" has lost any objective meaning, and is merely used as a kind of professional philosophical invective. [20, p. 363]

So he uses a very different language from that of Feigl above.

To my eye, not a single one of the papers on particular empirical sciences in the proceedings of the 1960 Stanford congress [16] is written under the paradigm that Tarski has in mind above. From his remarks in 1941, I doubt that this would have surprised Tarski himself. And given the general usage of the word "methodology", it seems unlikely that Tarski would have expected many people outside a group of loyal followers to interpret the M in DLMPS in line with his own account of "the methodology of empirical sciences". So even a deference to Tarski would hardly give us reason to insist on keeping the M.

By contrast the word "logic" certainly does mark a major area within the scope of DLMPS. DLMPS Congresses continue to attract top quality speakers in all branches of mathematical logic. Two of the international members of DLMPS are specifically devoted to logic, and several national members have a particular interest in it. Since logic is not a subset of philosophy of science, or indeed of philosophy at all, it follows that as things are at present, there is no question of dropping the L from DLMPS.

But the world moves on. Around 1950 some logicians—Bocheński in particular [21]—wanted an affiliation of "logic" to ICSU in order to get a wider recognition for modern logic. In this they succeeded magnificently. But logic today gets incomparably more recognition from its role in computer science than it does from the title of DLMPS. Logicians now have so many international outlets that they depend on DLMPS much less than a few decades ago, and this trend will probably continue.

Also in 1955, mathematical logic had stronger links with foundations than it does today. E.g., mathematical model theory, which was still finding its feet in 1955, is now a branch of mathematics like any other; it has interesting foundations but it is not itself a contribution to foundations. So the links between mathematical logic and philosophy of science grow weaker.

There are already signs that mathematical (as opposed to philosophical) logic may eventually part company from DLMPS. The trend is for fewer papers in mathematical logic to be submitted to DLMPS congresses. It seems very likely that DLMPS congresses will continue to attract philosophical work that uses mathematical logic, but less of the straight mathematics will find its way there. The General Assembly in Nancy was the first one to which the *Association for Symbolic Logic* sent no delegates; this was certainly an unintended accident and not a policy decision, but there is a message in the accident.

My own reaction would be to let rivers find their own natural course. The L in DLMPS should be secure for some decades to come.

## Bibliography

[1] Stanislas Dockx, 'Note historique concernant la fondation de l'Union internationale de Philosophie des Sciences'. *Dialectica* 31 (1977) 35–38.

[2] Anita Burdman Feferman and Solomon Feferman, *Alfred Tarski: Life and Logic*, Cambridge University Press, Cambridge 2008.

[3] Herbert Feigl, 'Scientific method without metaphysical presuppositions', *Philosophical Studies* 5 (1954) 17–29.

[4] Ferdinand Gonseth, *Les Mathématiques et la Réalité: Essai sur la Méthode Axiomatique*, Alcan, Paris 1936.

[5] Ferdinand Gonseth, 'Remarque sur l'idée de complémentarité', *Dialectica* 2 (1948) 413–420.

[6] Ferdinand Gonseth, 'La question de la méthode en psychologie', *Dialectica* 3 (1949) 324–337.

[7] Ferdinand Gonseth, 'La méthodologie des sciences peut-elle être élevée au rang de discipline scientifique?', *Dialectica* 11 (1957) 9–20.

[8] Ferdinand Gonseth, *Le Problème du Temps: Essai sur la Méthodologie de la Recherche*, Griffon, Neuchâtel 1964.

[9] Ferdinand Gonseth, *Logique et Philosophie Mathématiques*, Hermann, Paris 1998.

[10] Frank Greenaway, *Science International: A history of the International Council of Scientific Unions*, Cambridge University Press, Cambridge 1996.

[11] Robert Halleux and Benoît Severyns, 'Twenty-five years of international institutions', *Llull: Revista de la Sociedad Española de Historia de las Ciencias* 26 (2003) 315–321.

[12] Gerhard Heinzmann, 'Paul Bernays et la philosophie ouverte', in: James Gasser and Henri Volken (eds.), *Logic and Set Theory in 20th Century Switzerland*, PhilSwiss Schriften zur Philosophie, Band 1, PhilSwiss, Bern 2001, pp. 19–29.

[13] Wilfrid Hodges, 'DLMPS-Tarski's vision and our own', in: Peter Schroeder-Heister, Gerhard Heinzmann, Wilfrid Hodges, and Pierre Edouard Bour (eds.), Logic, Methodology and Philosophy of Science, Logic and Science Facing the New Technologies, Proceedings of the 14th International Congress (Nancy), College Publications, London 2015, pp. 9–26.

[14] Olli Lehto, *Mathematics without Borders: A History of the International Mathematical Union*, Springer, New York 1998.

[15] Henry L. Mansel and John Veitch (eds.), *Lectures on Metaphysics and Logic by Sir William Halmilton, Bart. Vol. II. Logic.*, Gould & Lincoln, Boston 1860.

[16] Ernest Nagel, Patrick Suppes and Alfred Tarski eds., *Logic, Methodology and Philosophy of Science, Proceedings of the 1960 International Congress*, Stanford University Press, Stanford CA 1962.

[17] Henri Poincaré, *Science et Méthode*, Flammarion, Paris 1908.

[18] John Corcoran (ed.), *Logic, Semantics, Metamathematics. Papers from 1923 to 1938 by Alfred Tarski. Translated by J. H. Woodger*, 2nd edition Hackett Publishing Company, Indianapolis IN 1983.

[19] Alfred Tarski, *Introduction to Logic and to the Methodology of the Deductive Sciences*, Oxford University Press, New York 1941.

[20] Alfred Tarski, 'The semantic conception of truth: and the foundations of semantics', *Philosophy and Phenomenological Research* 4 (1944) 341–376.

[21] Paul van Ulsen, 'The birth pangs of DLMPS', in this volume.

# Gespräch mit Arnold Oberschelp, dem Vorsitzenden der DVMLG von 1970 bis 1976

## Deborah Kant[1,2] & Deniz Sarikaya[3]*

[1] Fachbereich Mathematik, Universität Hamburg, Bundesstraße 55, 20146 Hamburg, Deutschland
[2] Fachbereich Philosophie, Universität Konstanz, Universitätsstraße 10, 78464 Konstanz, Deutschland
[3] Centre for Logic and Philosophy of Science, Vrije Universiteit Brussel, Pleinlaan 2, 1050 Brussel, Belgien
E-Mail: deborah.kant@uni-hamburg.de, deniz.sarikaya@vub.be

Arnold Oberschelp studierte Mathematik und Physik in Göttingen und Münster. Im Jahre 1957 wurde er an der Westfälischen Wilhelms-Universität Münster mit einer Arbeit mit dem Titel *Über die Axiome produktabgeschlossener arithmetischer Klassen*, betreut von Hans Hermes, promoviert.[1] Anschließend ging er an die Technische Hochschule Hannover, wo er sich im Jahre 1961 in Mathematik habilitierte. Es folgte eine Gastprofessur an der *University of California at Berkeley* 1967/68; im Jahre 1968 folgte er einem Ruf an die Christian-Albrechts-Universität zu Kiel, wo er die Nachfolge des Beweistheoretikers Kurt Schütte antrat. Zunächst war sein Lehrstuhl dem philosophischen Seminar zugeordnet, später dann dem Institut für Logik an der Mathematisch-Naturwissenschaftlichen Fakultät, dessen Gründungsdirektor er war. Nach Oberschelps Emeritierung im Jahre 1997 wurde Otmar Spinas an das Mathematische Seminar in Kiel als Professor berufen, sodass die Kieler Logik seither fortbesteht.

Oberschelp wurde auf der Mitgliederversammlung der DVMLG am 16. April 1964 in Oberwolfach als neues Mitglied zugewählt[2], war von 1965 bis 1978 im Vorstand der DVMLG[3] und von 1970 bis 1976 Vorsitzender der DVMLG.[4]

Anlässlich des sechzigsten Jubiläums der DVMLG trafen sich Arnold Oberschelp und die beiden Autoren digital am 15. Februar 2022 zu einem Gespräch. Der Text in diesem Artikel ist keine wörtliche Transkription des Gesprächs, sondern ein aus zu verschiedenen Zeitpunkten des Gesprächs ge-

---

*Die Autoren danken dem Deutschen Institut für Normung und dem Mathematischen Forschungsinstitut Oberwolfach für die freundliche Genehmigung für die Verwendung des Bildmaterials.

[1]Vgl. A. Oberschelp, Über die Axiome produkt-abgeschlossener arithmetischer Klassen, Archiv für mathematische Logik und Grundlagenforschung 4:3-4 (1958), S. 95–123.

[2]Sitzung vom 16. April 1964; DVMLG-Archiv A27.

[3]Protokoll über die Mitgliederversammlung der DVMLG in Oberwolfach am 8. April 1965; DVMLG-Archiv A38.

[4]Aktennotiz zum Protokoll der Mitgliederversammlung der DVMLG vom 9. 4. 1970 in Oberwolfach v. 16. 4. 1970; DVMLG-Archiv A51.

ABBILDUNG 1. *Links.* Arnold Oberschelp bei der Emeritierung Heinrich Behnkes in Münster am 4. März 1967. *Mitte.* Arnold Oberschelp in Aachen, 1978. *Rechts.* Arnold Oberschelp in Oberwolfach, 1987. (Fotos: Konrad Jacobs. Quelle: Bildarchiv des Mathematischen Forschungsinstituts Oberwolfach.)

nannten Aspekten rekonstruierter Text.[5] Die Autoren danken Herrn Oberschelp für das informative Gespräch.

*D.K. & D.S.* Herr Oberschelp, Sie sind seit nunmehr 58 Jahren Mitglied der DVMLG und sind in der allerersten Sitzung des Vereins nach der Gründungsversammlung zugewählt worden. Wer waren die prägenden Persönlichkeiten der DVMLG in den ersten Jahren?[6]

*Oberschelp.* Die damaligen Logikprofessoren, z.B. [Hans] Hermes und [Kurt] Schütte haben den Verein belebt. In den sechziger Jahren gab es noch nicht viele Logiker an deutschen Universitäten. Der Verein war von Arnold Schmidt in Marburg mit sieben Gründungsmitgliedern gegründet worden. Gründungsmitglieder waren u.a. Arnold Schmidt, Hermes, [Gisbert] Hasenjaeger aus Münster und Schütte, der damals auch in Marburg war. Der Vorsitzende Arnold Schmidt war eine sehr schwierige Person, mit dem es leicht Streit gab. Um ihn in seine Schranken zu weisen, bin ich einmal von Oberwolfach in den Ort [Wolfach] gefahren, habe eine Taschenausgabe des Bürgerlichen Gesetzbuches gekauft, um mit dem Finger darauf zeigen zu können, dass die Mitglieder den Vorsitzenden bestimmen.

---

[5] Zusätzlich zum Gespräch am 15. Februar 2022 wurden Gesprächsnotizen eines Telefonats von Benedikt Löwe mit Herrn Oberschelp vom 9. April 2021 für diese Rekonstruktion verwendet.

[6] Vgl. B. Löwe, Die Mitgliederentwicklung in der Frühzeit der DVMLG, in diesem Bande.

*D.K. & D.S.* Wie Sie gerade andeuteten, organisierte die DVMLG in den ersten Jahren regelmäßige Tagungen am Mathematischen Forschungsinstitut Oberwolfach (MFO), der idyllisch im Schwarzwald am Fluss Wolf gelegenen Forschungsinstitution. Was sind Ihre prägenden Erinnerungen an diese Tagungen?

*Oberschelp.* Die ersten Jahre waren eine sehr schöne Zeit. Ich bin jedes Jahr einmal zur DVMLG-Tagung nach Oberwolfach gefahren, wo man sich mit den anderen Mitgliedern traf, meistens im Frühjahr, um die mathematische Logik weiterzuentwickeln. Jeder, der im vergangenen Jahr Forschungsfortschritt gemacht hatte, berichtete in Oberwolfach darüber. Themen waren z.B. Modelle der Berechenbarkeit. Wir lernten die jungen Schweizer Logiker kennen und im Jahre 1967 kam auch [Paul] Bernays aus der Schweiz als großer Besuch auf unsere Tagung.[7]

*D.K. & D.S.* Das MFO war und ist eine mathematische Forschungseinrichtung und die Tatsache, dass in den frühen Jahren alle Mitgliederversammlungen am MFO stattfanden, kann als starke Ausrichtung zur Mathematik verstanden werden. War die DVMLG in den frühen Jahren hauptsächlich ein Verein von Mathematikern?

*Oberschelp.* Die meisten Mitglieder des Vereins waren in der Mathematik zu Hause und es gab auch einige Mitglieder, die der Philosophie außerordentlich feindlich gegenüberstanden.[8] Ich erinnere mich an einen ernsten Streit zwischen den Mathematikern und den Philosophen unter den Logikern an einer deutschen Universität, in dem es um einen Habilitanden aus der Philosophie ging, der von einem Mathematiker blockiert wurde.

In der DVMLG gab es auch Debatten um den Namen des Vereins. Viele Mathematiker unter den Mitgliedern wollten den mathematischen Charakter des Vereins hervorheben und sahen ihn als einen Verein für mathematische Logik. Auf der anderen Seite gab es auch Philosophen unter den Mitgliedern, die mit der Beschränkung auf die mathematische Logik nicht zufrieden waren und gerne das Wort „mathematisch" aus dem Namen des Vereins entfernt hätten. Aber die Mehrzahl der Mitglieder waren im Herzen Mathematiker. Auf der anderen Seite hielt man in der weiteren Welt der Mathematik oft nicht viel von der mathematischen Logik.

*D.K. & D.S.* Hat diese Spannung zwischen mathematischer Logik und Philosophie Ihren eigenen Weg beeinflusst?

---

[7]Diese Bemerkung bezieht sich auf die Tagung *Zur mathematischen Logik, 3.–8. April 1967*; vgl. Gästebuch 2 (08. Juni 1954 – 13. Juni 1970), Oberwolfach Digital Archive E20/00149, Blatt 227.

[8]Vgl. hierzu B. Löwe, Grundlagenforschung der exakten Wissenschaften: die DVMLG und die Philosophie, in diesem Bande, § 4.

ABBILDUNG 2. Arnold Oberschelp wird am 26. September 2019 die Beuth-Denkmünze von Daniel Schmidt, Mitglied der Geschäftsleitung von DIN, verliehen (Quelle: DIN e.V.)

*Oberschelp.* Ich kam als Assistent nach Hannover in ein Umfeld mit sehr aufgeschlossenen Personen, die nicht fanatisch waren. Als ich dann nach Kiel kam, war ich erst in der Philosophie einer von drei Professoren: die anderen waren Kurt Hübner und Hermann Schmitz. Später leitete ich dann mein eigenes kleines Institut. Die skeptische Haltung der Mathematiker gegenüber der Logik habe ich in Kiel dann auch bemerkt: ich saß zwischen den Stühlen und war eher am Rande der Mathematik. Ich hatte Herrn [Klaus] Potthoff aus Hannover mitgebracht und zusammen mit ihm und meinen Assistenten bildete die Kieler Logik eine sehr schöne Gemeinschaft. Gemeinsam haben wir 1974 das *Logic Colloquium* nach Kiel geholt, zu dem auch die internationale Prominenz der Logik kam.[9]

---

[9]Vgl. G. H. Müller, A. Oberschelp, K. Potthoff, Hrsgg., $\models$ISILC Logic Conference. Proceedings of the International Summer Institute and Logic Colloquium, Kiel 1974, Lecture Notes in Mathematics, Band 499, Springer-Verlag 1975.

Im Kontrast dazu hat die Jahrestagung der Deutschen Mathematiker-Vereinigung noch nie in Kiel stattgefunden und die Universität Kiel hat auch, im Gegensatz zu anderen Universitäten, keinen DMV-Ansprechpartner. Die Kieler Logik hingegen hat sich immer bemüht, Tagungen nach Kiel zu holen. Herr Potthoff hat übrigens inzwischen die Logik verlassen und hat sich dem Weinhandel gewidmet.[10]

*D.K. & D.S.* Das erwähnte Kieler *logic Colloquium* fiel in Ihre Zeit als Vorsitzender der DVMLG. Hatten Sie in der Zeit noch andere Erfahrungen mit Internationalisierung?

*Oberschelp.* In meiner Eigenschaft als Vorsitzender vertrat ich die Logik der Bundesrepublik auf Kongressen. Dort traf man dann unter anderem die Kollegen aus Ost-Deutschland, insbesondere Ost-Berlin. Die innerdeutschen Spannungen hatten auch Auswirkungen auf die Logik: ein Berliner Kollege hatte auf einen Ruf nach Münster gehofft, den dann aber Hermes bekommen hat und [Wolfgang] Schwabhäuser hat es geschafft, einen Israelbesuch zu nutzen, um aus der DDR zu fliehen und nach West-Deutschland zu kommen.

In den Jahren der Stellenknappheit sind viele Logiker nach Amerika gegangen: der größere Markt erlaubte es Nachwuchswissenschaftlern, eine feste Anstellung zu bekommen. So z.B. ein Assistent von Hermes, Hubert Schneider, der an die *University of Nebraska* gegangen ist. Hubert Schneider ist dann später leider bei einem Autounfall im mittleren Westen ums Leben gekommen. Ich selbst habe auch eine Zeit an der *University of California in Berkeley* als Gastprofessor verbracht.

*D.K. & D.S.* Wie gestaltete sich die Übergabe des Vorsitzes der DVMLG und die Zeit danach?

*Oberschelp.* Ich hatte etwas Mühe, den Vorsitz loszuwerden. Ich habe deutlich gesagt, dass ich genug für den Verein getan habe, aber es hat dann doch länger gedauert, bis sich ein Nachfolger gefunden hat. Ich habe mich danach den Aufgaben im Ausschuss für Einheiten und Formelgrößen des DIN gewidmet. Es gab allerdings keine Verbindung zwischen diesen beiden Aufgaben: der Kontakt lief über einen Hannoveraner Kollegen.[11]

---

[10]Klaus Potthoff ist seit 1982 Inhaber der Fair Wein GmbH in Boksee bei Kiel und war von 1985 bis 1990 Vorsitzender der DVMLG.

[11]Oberschelp bezieht sich auf Wilhelm Quade (1898–1975), der langjähriges Mitglied des Ausschusses für Einheiten und Formelgrößen (AEF) im Deutschen Institut für Normung e.V. (DIN) war; vgl. Stefan Schottlaender, Wilhelm Quade, *1.12.1898 †10.6.1975, Nachruf der Braunschweigischen Wissenschaftlichen Gesellschaft, vorgetragen in der Plenarsitzung am 8.12.1978, Abhandlungen der Braunschweigischen Wissenschaftlichen Gesellschaft 30 (1979), S. 145–148.

Im Jahre 1976 übernahm Gert H. Müller den Vorsitz der DVMLG von Arnold Oberschelp für die nächsten fünf Jahre. Oberschelp wurde vom AEF (vgl. Anm. 11; jetzt *Fachbereich 1 „Einheiten und Formelgrößen" des DIN-Normenausschusses Technische Grundlagen*) um seine Expertise bei der Erarbeitung von Normen über Zeichen der Mengenlehre (DIN 5473) und der mathematischen Logik (DIN 5474) gebeten. Später arbeitete er auch an der Überarbeitung der Norm über mathematische Zeichen (DIN 1302), der Norm über physikalische Größen (DIN 1313) und an anderen Projekten mit. Für dieses Engagement wurde ihm 2019 die Beuth-Denkmünze verliehen (vgl. Abb. 2).

# Unterwegs mit Alan Turing

## Anke Kell

Ohnsorg-Theater, Heidi-Kabel-Platz 1, 20099 Hamburg, Deutschland

Manchmal gehen einer Geschichte nackte Zahlen voraus. So war es auch bei der ersten großen Gastspielreise in der Geschichte des *University Players* e.V., der englischsprachigen Theatergruppe am Institut für Anglistik und Amerikanistik an der Universität Hamburg. Die wichtigste Zahl, die ich diesem Beitrag voranstellen möchte, ist: zehn. Etwas mehr als zehn Jahre ist es her, seit Benedikt Löwe von der Universität Hamburg mich kontaktierte, um den *University Players* eine mobile Produktion des Theaterstückes *Breaking the Code* (1986) von Hugh Whitemore vorzuschlagen—anlässlich des 100. Geburtstages von Alan Turing.

Diese Gastspielreise und die Begegnung mit Menschen aus vollkommen anderen Kontexten bedeutete viel für alle Beteiligten und prägte uns nachhaltig. Da ich im Moment keinen Zugriff auf die Dokumentation habe, verlasse ich mich bei der Rekonstruktion des Reiseabenteuers auf mein Gedächtnis und den Briefwechsel mit Benedikt Löwe, den er für mich freundlicherweise noch einmal exzerpiert hat. Von 2009 bis 2013 war ich als Produktionsleiterin der *University Players* und Wissenschaftliche Mitarbeiterin fest an der Universität Hamburg angestellt. Seit 2013 arbeite ich nun als Dramaturgin an verschiedenen Theatern. Mein Umfeld hat sich also stark verändert, auch wenn „Theater" eine Konstante geblieben ist.

„*To begin at the beginning*", um mit Dylan Thomas zu sprechen: Anlässlich besagten Geburtstages des bedeutenden Computerpioniers Alan Turing sollten im Jahre 2012 zahlreiche internationale Konferenzen und Tagungen in der gesamten Welt stattfinden, ausgerichtet vom *Turing Centenary Advisory Committee*, geleitet von Barry Cooper aus Leeds. Es gab einige nationale Unterkomitees, unter anderem das deutsche Unterkomitee, geleitet von Benedikt Löwe.

Das Theaterstück *Breaking the Code* basiert auf der Alan Turing-Biografie *The Enigma* (1983) von Andrew Hodges. Es porträtiert Person und Arbeit des genialen britischen Mathematikers, der an den gesellschaftlichen Ressentiments gegenüber seiner Homosexualität zerbricht. Der Autor Hugh Whitemore unternimmt in dem Stück den Versuch, die Forschungsarbeit des Wissenschaftlers in Relation zu dessen Persönlichkeit zu setzen.

Als Ergänzung und Bereicherung wissenschaftlicher Konferenzen zu Ehren Alan Turings schien sich das Stück gut zu eignen, auch wenn die sehr unterschiedlichen Aufführungsorte, von denen die meisten wenig Ähnlichkeit mit Theaterräumen bzw. -bühnen hatten, für uns als nicht professionelle Gruppe keine kleine Herausforderung darstellen sollte. Kurzum: Es schien

ABBILDUNG 1. *Von links nach rechts.* Jeff Caster als Mick Ross, Maximilian Duchow as Alan Turing und Jonathan Guss als Dillwyn Knox bei einer Probe im Hamburger LICHTHOF Theater am 16. September 2012. (Bilder: Steffen Baraniak. Copyright: G2 Baraniak.)

ein interessantes Unterfangen zu sein, aber ob es realisierbar wäre, blieb zunächst offen.

Ich befand mich zu diesem Zeitpunkt in der „zweiten Spielzeit" bzw. im vierten Semester meiner Tätigkeit für den *Theatre Workshop*, der seit 1984 eine feste Institution an der Universität Hamburg ist. Normalerweise produzieren die *University Players* mit Studierenden Theaterstücke, die jeweils am Ende eines Semesters im Audimax aufgeführt werden. Mit *Breaking the Code* wurde nun eine zusätzliche Sonderproduktion angefragt, für die es erst einmal keine Infrastruktur und—viel entscheidender—kein Budget gab.

Eigentlich hielten wir kaum etwas Konkretes zu Beginn unseres Projektes in Händen, das zum ersten Mal auch die Fachbereiche *Sprache, Literatur, Medien* und *Mathematik* miteinander in Kontakt brachte. Es existierte lediglich die Idee, die Abenteuerlust eines recht heterogenen und flexiblen Vereins und ein bisschen Zeit, Parameter abzustecken und Realisierbarkeit zu prüfen.

Zeit schien dann die kostbarste Ressource in Vorbereitung auf unsere Unternehmung zu sein, denn Verlag, Agenten, Künstler und Sponsoren ließen sich nicht unmittelbar begeistern, sondern bedurften sensibler Überzeugungsarbeit, was vor allem mit dem mobilen Charakter unseres Projektes zusammenhing.

Aber—wie es bei Shakespeare heißt—„*all's well that ends well*": Etwa ein halbes Jahr nach der Anfrage nahm unsere Produktion Konturen an: Sponsoren waren gefunden, die Aufführungsrechte gesichert, die Spielorte

ABBILDUNG 2. *Links.* Koffer als Sinnbild der Heimatlosigkeit. *Rechts.* Kostüme als zeitliche und zeitlose Verortung (von links nach rechts: Matthias Maurer, Nora Farrell, Saskia Wieland, Jeff Caster). Aufnahmen einer Probe im Hamburger LICHTHOF Theater am 16. September 2012. (Bilder: Steffen Baraniak. Copyright: G2 Baraniak.)

gebucht und das künstlerische Basisteam begann mit ersten Konzeptionsideen.

Mit Jeff Caster konnten wir einen erfahrenen Regisseur und Schauspieler gewinnen, der sich schon seit vielen Jahren um die *University Players* und andere englischsprachige Theatergruppen aus dem Amateur- und Profibereich verdient gemacht hatte. Nach und nach war es uns möglich, auch die anderen relevanten künstlerischen und technischen Positionen mit Mitgliedern der *University Players* zu besetzen, die sich für eine Produktion „*on the road*" engagieren wollten. Dazu sei erwähnt, dass die *University Players* nahezu ausschließlich ehrenamtlich tätig sind. Es existiert lediglich die feste (halbe) Stelle der Produktionsleitung am Institut für Anglistik und Amerikanistik—aber damit allein lässt sich kein Theater machen. Die einzelnen Gewerke, die in einem Theaterstück zusammenwirken, werden bei den *University Players* für jede Produktion neu gesucht und besetzt.

Waren wir es bislang gewohnt, an festen Bühnen zu spielen, verwandelten wir uns für *Breaking the Code* nach alter Gaukler-Tradition zum fahrenden Volk, was unsere kreativen Energien besonders beflügelte. So ersann die Bühnenbildnerin Julie Junge einen mobilen Spielraum, der ausschließlich aus Koffern bestand, die Schränke und Sitzgelegenheiten markierten und auf inhaltlicher Ebene die Heimatlosigkeit von Alan Turing verdeutlichten.

ABBILDUNG 3. *Links.* Von links nach rechts: Nora Farrell, Saskia Wieland, Jonathan Guss, Maximilian Duchow, Matthias Maurer. *Rechts.* Nora Farrell als Sara Turing und Jonathan Guss als Dillwyn Knox. Aufnahmen einer Probe im Hamburger LICHTHOF Theater am 16. September 2012. (Bilder: Steffen Baraniak. Copyright: G2 Baraniak.)

Teresa Musal konzipierte Kostüme, die zwei Ebenen bedienten: Sie verorteten die Figuren im England der 1950er Jahre, der Zeit, in der die Geschichte spielt, und kennzeichneten zugleich die Zeitlosigkeit der Geschichte eines Menschen, der lediglich als Instrument dient und dessen Persönlichkeit im Verborgenen bleiben soll. Praktischerweise passten diese Kostüme wunderbar in die Koffer, so dass wir als große Reisegruppe mit ausladendem Reisegepäck unkompliziert mit der Deutschen Bahn zu den verschiedenen Spielorten fahren konnten. Normalerweise kommen Gastspielreisen von Theatern nicht ohne Transporter—meistens eher LKW—aus, um Kulissen, Requisiten, Technik und alles, was zu einer Aufführung gehört, von einer Bühne zur anderen zu transferieren (vgl. Abbildung 2 rechts).

Nach einer mehrwöchigen Probenphase im Sommer 2012 nahm die Tournee ihren Anfang: Unsere Premiere spielten und feierten wir am 14. September 2012 im Heinz Nixdorf MuseumsForum in Paderborn anlässlich des Colloquium Logicum auf großer Bühne. Bereits zwei Tage später konnten wir für ein Hamburger Publikum im LICHTHOF Theater spielen.

Wiederum drei Tage später gaben sich Realität und Fiktion ein Stelldichein—denn wir spielten in einem Hörsaal der TU Braunschweig im Rahmen der Jahrestagung der *Gesellschaft für Informatik* e.V. So nah wie dort kamen sich Wissenschaft und Kunst selten, versucht doch auch das Theaterstück, beide Ebenen miteinander zu verflechten.

Nach diesen ersten drei Vorstellungen in unterschiedlichsten Räumen, für die wir jedes Mal individuelle Gestaltungsideen entwickelten (man bedenke nur die verschiedenen Beleuchtungsmöglichkeiten—und Licht ist ein essenzielles Element in Theaterinszenierungen), konnten wir wertvolle Erfahrungen sammeln, um gerüstet zu sein für die nächste Herausforderung: Amsterdam.

Erneut setzten wir uns mitsamt Ausstattung in Bewegung. Wir müssen auf den Bahnsteigen ein interessantes Bild abgegeben haben: Zehn meist junge Leute unter dreißig Jahren mit alten, verbeulten Koffern, von denen einer so groß war, dass er einen eigenen Sitzplatz brauchte. Jedes Mal bangten wir ein wenig, ob das Bahnpersonal wohl Einspruch gegen unseren sperrigen Reisekumpanen erheben würde. Wir hatten Glück: Bis auf eine längere Diskussion mit einem sehr korrekten Beamten blieben uns Probleme erspart.

In Amsterdam waren Anfang Oktober 2012 im Rahmen der Jahrestagung der *Nederlandse Vereniging voor Logica en Wijsbegeerte der exacte Wetenschappen* und des *Open Dag* am *Science Park Amsterdam* gleich zwei Vorstellungen hintereinander in verschiedenen Räumlichkeiten geplant: eine im *Eetcafé Oerknal* im *Science Park Amsterdam*, einem gemütlichen studentischen Bistro, und eine im *Hella Haasse en Simon Vestdijkzaal* in der *Openbare Bibliothek Amsterdam*. Während ersterer unsere Improvisationsfreude entfachte, konnte letzterer mit allen Vorteilen eines Theatersaals aufwarten, was zu zwei sehr unterschiedlichen und besonderen Vorstellungen führte.

Amsterdam blieb uns allen in spezieller Erinnerung, weil wir zum ersten Mal seit Beginn der Tournee mehr Zeit an einem Ort verbringen konnten—und der Ablauf von Ankunft, Aufbau, Vorstellung, Abbau angenehm durchbrochen wurde. Zudem nutzten wir die Gelegenheit, die Reize der aufregenden Stadt zu genießen.

Wir waren als Team mittlerweile gut zusammengewachsen, so dass unsere letzte Vorstellung in Almere etwa einen Monat später fast schon Routine war. An diese Stadt habe ich fast die lebendigste Erinnerung, weil sie sich so stark von den anderen unterschied.

Almere wurde erst 1975 nur 25 km östlich von Amsterdam auf einem trocken gelegten Gebiet erbaut. Nichts in dieser Stadt ist demnach älter als 47 Jahre—also nur drei Jahre älter als ich selbst. Die Charakteristik eines künstlichen Gesamtentwurfs vermittelt sich dort überall, die Stadt wirkt wie ein Konstrukt, das auf seine Patina wartet. Mitten im Zentrum befindet sich die imposante Schouwburg Almere, in deren Kleinen Saal wir anlässlich der Jahreskonferenz des dortigen *Alan Turing Institute* spielen durften.

So endete die erste Tournee der *University Players*, die für uns alle eine aufregende und lehrreiche Erfahrung war. Dank der wagemutigen Idee von Benedikt Löwe, eine nicht professionelle studentische Theatergruppe als kulturelle und künstlerische Bereicherung für wissenschaftliche Konferenzen zu engagieren und seines unermüdlichen Einsatzes für externe Stiftungs- und Sponsoren-Gelder war uns dieses Abenteuer vergönnt.

ABBILDUNG 4. *Links.* Jeff Caster als Mick Ross und Maximilian Duchow als Alan Turing. *Rechts.* Maximilian Duchow als Alan Turing, Jonathan Guss als Dillwyn Knox, Saskia Wieland als Pat Green und Nora Farrell als Sara Turing. Aufnahmen einer Probe im Hamburger LICHTHOF Theater am 16. September 2012. (Bilder: Steffen Baraniak. Copyright: G2 Baraniak.)

Daher freue ich mich, nach zehn Jahren hier noch einmal die Gelegenheit nutzen zu können, um mich bei allen Mitstreitern dieses Projekts zu bedanken, namentlich bei den Schauspielern Maximilian Josef Duchow, Matthias Maurer, Nora Farrell, Saskia Wieland und Jonathan Guss, den Technikern Björn Mahrt, Anna Determann und Paul-Louis Lelièvre, der Bühnenbildnerin Julie Junge, der Kostümbildnerin Teresa Musal, der Produktionsassistentin Marie-Caroline Schulte—und nicht zuletzt: bei Jeff Caster, der sowohl die Regie als auch eine Rolle übernahm. Großer Dank gebührt zudem der Heinz Nixdorf Stiftung Westfalen, der Deutschen Vereinigung für Mathematische Logik und für Grundlagenforschung der exakten Wissenschaften und der Universität Hamburg, die diese außergewöhnliche Theaterreise durch ihre Finanzierung möglich gemacht haben.

Beenden möchte ich diesen Beitrag mit einer sentimentalen Note: Jeff Caster war zwar nicht der erste Regisseur, den ich für *Breaking the Code* angefragt habe, aber der einzige, der sich die Inszenierung, die vielen äußeren Umständen gerecht werden musste, zutraute—sowohl zeitlich als auch künstlerisch. Man muss es so sagen, wie es ist: Regie ist die anspruchsvollste, anstrengendste und verantwortungsvollste Aufgabe im Theater. Wenn ich an *Breaking the Code* denke, kommt mir als erstes Jeffs grenzenlose Energie und Unerschütterlichkeit in den Sinn. Leider ist er am 3. Januar 2015 in Vietnam an einer Infektion gestorben.

# Mathematical logic at the Department of Mathematics at the TU Darmstadt

Ulrich Kohlenbach & Thomas Streicher

Department of Mathematics, Technische Universität Darmstadt, Schlossgartenstraße 7, 64289 Darmstadt, Germany
E-mail: {kohlenbach,streicher}@mathematik.tu-darmstadt.de

The history of logic in Darmstadt goes back to 1874, when Ernst Schröder (1841–1902) was appointed as a professor at the *Polytechnikum Darmstadt* (1874–1876) which in 1877 became the *Technische Hochschule Darmstadt*.

The precursor of the current research group in logic was created in the year 1970 when Rudolf Wille (1937–2017) founded the research group *AG1: Allgemeine Algebra*. It was enlarged in 1971 with the appointments of Peter Burmeister (1941–2019) and Klaus Keimel (1939–2017) as new professors.[1] Later, Christian Herrmann (born 1943), who had received his doctorate under the supervision of Wille in 1972, became an *außerplanmäßiger Professor* (extraordinary professor) in this group.

In addition to universal algebra, topics of AG1 were partial and topological algebraic structures, discrete mathematics, and lattice theory involving aspects of decidability and axiomatizability.

Also in 1971, Peter Zahn (born 1930) came to Darmstadt joining the research group of Detlef Laugwitz, where he completed his Habilitation in 1973 and later became *außerplanmäßiger Professor*; after Laugwitz's retirement, Zahn joined the logic group working primarily on predicative foundations of mathematics influenced by Paul Lorenzen. In 1978, Bernhard Ganter (born 1949), who had received his doctorate under the supervision of Rudolf Wille in 1974, joined the AG1 as professor but moved in 1993 to the TU Dresden.

The research group on *Formal Concept Analysis* emerged from AG1; this group focused on graph-based logic systems for concept analysis in knowledge acquisition and processing applications (Burmeister, Ganter, Wille). Research in this direction is still being pursued in co-operation with the *Ernst Schröder Zentrum für Begriffliche Wissensverarbeitung* e.V. which started in 1983 with Wille as its first *Sprecher*.

During the second half of the 1970s Darmstadt was one site of the multinational seminar on Dana Scott's continuous lattices resulting in the famous *Compendium of Continuous Lattices* published with Springer in 1980. The Darmstadt site was represented by Klaus Keimel and his student Gerhard

---
[1] Logic has been also represented at the Computer Science Department, notably by Wolfgang Bibel (born 1938), one of the pioneers of Artificial Intelligence in Germany, who founded the research group of *Intellektik* (1988–2004).

Gierz who also worked on sheaf representations of ordered algebraic structures. Karl-Heinrich Hofmann (born 1932), another co-author of this volume, joined the Mathematics Department in Darmstadt in 1982.

Keimel was also the main contributor and organizer of the later edition *Continuous Lattices and Domains* which appeared in 2003 with Cambridge University Press. This was a strongly revised version which systematically treated also the more general case of domains, i.e., directed complete partial orders lacking a top element, which are of central importance in the denotational semantics of programming languages.

One may add here that Gerd Mitschke worked in Darmstadt during the first half of the 1970s and organized in Darmstadt one of the first informal meetings on $\lambda$-calculus (20–25 August 1973) gathering the few experts who initiated the renaissance of this subject including Henk Barendregt, Roger Hindley, Gordon Plotkin and Chris Wadsworth. Mitschke's work was mainly syntactical and disjoint from the group of people interested in continuous lattices.

In 1995 Klaus Keimel, the newly appointed Thomas Streicher (born 1958) and the aforementioned Christian Herrmann started the *AG14: Logik und mathematische Grundlagen der Informatik* (Logic and the Mathematical Foundations of Computer Science) while the remaining AG1 focused on *Universal Algebra* (later joined by Thomas Ihringer, 1953–2015, who became *außerplanmäßiger Professor* in 1996) and *Formal Concept Analysis* (Burmeister, Wille).

Martin Hofmann (1965–2018), who together with Streicher introduced the groupoid model for Martin-Löf type theory, worked in the AG14 as research assistant from 1995 to 1998 and received his Habilitation in 1999 when he was already lecturer at the University of Edinburgh. He also joined the AG14 as a professor in the summer term 2001 before he moved to the Department of Computer Science of the *Ludwig-Maximilians-Universität München*. In this period starting with his habilitation work, he developed his type-theory based approach to characterize computational complexity classes in a logical way.

Both the AG1 and the AG14 were re-united around 2003/2004 in connection with the new appointments of Martin Otto (born 1961) in 2003 and—as successor of Rudolf Wille—Ulrich Kohlenbach (born 1962) in 2004 and the research group was renamed in *AG Logik*.

The logic group organized the *Colloquium Logicum 2008* (an installment of the biannual conference coordinated by the DVMLG) at TU Darmstadt with a special evening lecture by Professor Georg Kreisel in connection with his 85th birthday.

From 2010 to 2015, Martin Ziegler (born 1968, now Professor at KAIST) joined the AG Logik. With Kohlenbach and Ziegler as principal investiga-

| PhD candidate | Year | Supervisor |
| --- | --- | --- |
| Julian Bitterlich | 2019 | Otto |
| Eyvind Martol Briseid | 2009 | Kohlenbach |
| Felix Canavoi | 2018 | Otto |
| Jaime Gaspar | 2011 | Kohlenbach |
| Alexander Kartzow | 2011 | Otto |
| Daniel Körnlein | 2016 | Kohlenbach |
| Angeliki Koutsoukou-Argyraki | 2016 | Kohlenbach |
| Alexander P. Kreuzer | 2012 | Kohlenbach |
| Peter Lietz | 2004 | Streicher |
| Tobias Löw | 2006 | Streicher |
| Carsten Rösnick | 2014 | Ziegler |
| Pavol Safarik | 2013 | Kohlenbach |
| Florian Steinberg | 2016 | Kohlenbach & Ziegler |
| Jonathan Weinberger | 2021 | Streicher |

FIGURE 1. PhD graduations of the *AG Logik* in alphabetic order

tors, the *AG Logik* took part in the Darmstadt-Tokyo PhD School on Mathematical Fluid Dynamics (2011–2016, DFG International Research Training Group 1529). Since 2017, Kord Eickmeyer (born 1979) has been a permanent lecturer in the logic group. In 2021, Pascal Schweitzer, who had been appointed as professor in the *AG Didaktik* (Research Group: Mathematics Education), became also a co-member of the *AG Logik* given the close connection between his research and the area of Logic in Computer Science. Finally, in 2021, Anton Freund (born 1990), who had been a postdoctoral researcher in Kohlenbach's research group, received a prestigious *Emmy-Noether-Programm* award from the German Science Foundation (DFG) and was appointed in October 2021 as *Assistenzprofessor* in the *AG Logik*.

In recent years, the *AG Logik* has been representing the subject area of mathematical logic viewed as an applied foundational discipline between mathematics and computer science. Research activities focus on the application of proof-theoretic, recursion-theoretic, categorical, algebraic and model-theoretic methods from mathematical logic to mathematics and computer science. Besides classical mathematical logic (represented by proof theory, computability theory and model theory) this involves proof mining, constructive type theory, categorical logic, universal algebra, domain and lattice theory, finite model theory and complexity theory. Within mathematics, a primary field of applications in the proof- and recursion-theoretic setting is the extraction of new information from proofs in areas of core

| Habilitation candidate | Year | Mentor |
|---|---|---|
| Achim Blumensath | 2008 | Otto |
| Kord Eickmeyer | 2020 | Otto |
| Laurenţiu Leuştean | 2009 | Kohlenbach |
| Vassilis Gregoriades | 2015 | Kohlenbach |
| Sam Sanders | 2022 | Kohlenbach |
| Matthias Schröder | 2016 | Streicher |
| Benno van den Berg | 2011 | Streicher |

FIGURE 2. Habilitations of the Logic Group in alphabetic order

mathematics (proof mining: Kohlenbach). The goal of this applied reorientation of proof theory is the use of proof-theoretic transformations such as appropriate functional interpretations for the analysis of prima facie non-effective proofs in mathematics for the purpose of extracting new results from given proofs. This concerns qualitative aspects such as generalizations of proofs (e.g., from a linear Banach space setting to metric structures) and new uniformity results (independence of existence assertions from certain parameters) as well as quantitative aspects such as the extraction of explicit rates of convergence or—in cases where this is precluded—rates of metastability, oscillation bounds and other effective data from proofs. This novel proof mining approach, for which the logic group in Darmstadt is internationally recognized as the leading center, has been applied in many areas of (mostly nonlinear) analysis including approximation theory, fixed point theory, ergodic theory, abstract Cauchy problems, non-smooth optimization, hyperbolic geometry, pursuit-evasion games.

Complementing this applied proof-theoretic research, also foundational goals are pursued such as the calibration of the proof-theoretic strength of mathematical theorems, e.g., in combinatorics and concrete independence results (Freund). The proof-theoretic research is also connected to methods in categorical logic, constructive systems of set theory and type theory and homotopic type-theoretic foundations (Streicher) as well as dilators and ordinal notation systems (Freund).

The model-theoretic research of the logic group has close links to discrete mathematics (graphs and hypergraphs, Eickmeyer, Otto, Schweitzer) and algebra (group theory, Schweitzer).

Concerning logic in computer science and the mathematical foundations of computer science, major activities revolve around issues of semantics. On the one hand, this involves the mathematical foundation of the semantics and the logic of programming languages (Streicher); on the other hand, logics and formal systems are investigated in the sense of model theoretic semantics, with respect to expressiveness and definability, with an emphasis on computational aspects (algorithmic and finite model theory, descriptive complexity: Eickmeyer, Otto, Schweitzer) as well as practical and theoretical aspects of the graph isomorphism problem and algorithmic symmetry detection. Besides specific application domains in computer science, as, e.g., verification, data bases and knowledge representation, there is work on foundational issues in the areas of computability and complexity, as well as type theory and category theory (Streicher).

Overall, the logic group forms an internationally well connected cluster of expertise, with a characteristic emphasis on the connections that mathematical logic has to offer, both with respect to other areas within mathematics and with respect to logic in computer science.

Since the formation of the *AG Logik* in its current form in 2004 many doctoral dissertations and habilitations have been completed, listed in Figures 1 and 2.

The logic group conducted many research projects with external funding such as individual DFG projects in the areas of finite model theory (Blumensath, Otto), proof mining (Kohlenbach) and semantics (Keimel), participated in DFG-cooperation projects with, e.g., Novosibirsk (Herrmann, Keimel, Kohlenbach, Streicher) and South Africa (Keimel, Kohlenbach, Streicher) and took part in the EU working group APPSEM II (Keimel, Kohlenbach, Streicher, Ziegler).

Current research projects are *Continuous Order Transformations: A Bridge between Ordinal Analysis, Reverse Mathematics, and Combinatorics* (Emmy-Noether-Programm; Freund), *Proof Mining in Convex Optimization and related areas* (DFG KO 1737/6-2; Kohlenbach), and *Next generation algorithms for grabbing and exploiting symmetry* (ENGAGES, ERC Consolidator Grant; Schweitzer). The Logic Group takes also part in the new Research Profile Theme *Cognitive Science* of the TU Darmstadt (Kohlenbach).

# What can formal systems do for mathematics? A discussion through the lens of proof assistants

Angeliki Koutsoukou-Argyraki*

Department of Computer Science and Technology (Computer Laboratory), University of Cambridge, J. J. Thomson Avenue, Cambridge CB3 0FD, England
E-mail: ak2110@cam.ac.uk

### Abstract

This article containing interviews with Jeremy Avigad, Jasmin Blanchette, Frédéric Blanqui, Kevin Buzzard, Johan Commelin, Manuel Eberl, Timothy Gowers, Peter Koepke, Assia Mahboubi, Ursula Martin, and Lawrence C. Paulson attempts to approach the question of the significance of proof assistants—in tandem with the (possible) effects of their underlying logical formalisms—for contemporary and future mathematical practice. The answer to this broad question within such a fast-developing area involving cutting-edge research cannot be clear nor complete; here, through a discussion with eleven leading experts and considering some recent advances as well as several newly started research projects, we merely attempt to illuminate varying aspects of the topic, with the hope to demonstrate its dynamic, richness and potential.

## 1 Introduction

### 1.1 Formal systems and mathematics

The typical mathematician doing research in pure or applied mathematics is hardly ever concerned with foundations. Areas of mathematical logic, such as proof theory, homotopy type theory, model theory, set theory, computability theory/recursion theory, reverse mathematics, are often considered to be closer to, or classified under computer science instead of mathematics, even though they do have interesting applications in mainstream mathematics as well. E.g., in Kohlenbach's proof mining [51, 52, 53], a research program in applied proof theory, if certain prerequisites referring to the logical form of the statement proved and the logical framework at hand are fulfilled, certain logical metatheorems guarantee the pen-and-paper extractability of effective information even from nonconstructive proofs. Proof

---
*The author would like to warmly thank all participants for their valuable insights and comments and Benedikt Löwe for the paper invitation. She was supported by the ERC Advanced Grant ALEXANDRIA (Project GA 742178) led by Lawrence C. Paulson.

mining has found many applications over the last two decades mainly within nonlinear analysis (e.g. there are many results in Banach spaces, fixed point theory, convex optimisation, ergodic theory, topological dynamics and more).

All along the foundational crisis in mathematics fuelled by the paradoxes of the early 20th century, the efforts by Whitehead and Russell to axiomatise mathematics and express it in symbolic logic [80], the rise and fall (due to Gödel's Incompleteness Theorems) of Hilbert's Program and all the way until today, the typical mathematician working in subfields of e.g. algebra, geometry, topology, combinatorics, number theory or analysis (even more so a mathematician doing applied mathematics) would (almost) safely ignore foundations; even constructivism and intuitionism remain to most mathematicians a foreign land.

The proof that the continuum hypothesis is independent of ZFC by the combined works of Gödel and Cohen in 1963 was followed by the proofs of the independence of a number of statements mainly within the realm of logic and set theory. In the 1992 novel "Uncle Petros and Goldbach's Conjecture" by Doxiadis [28], the main character, a reclusive former mathematician who spent his youth, talent and potential in an obsessive, fruitless attempt to prove Goldbach's conjecture, desperately tries to find consolation (or to save himself from insanity) in the belief that his inability to prove Goldbach's conjecture despite his obsessive efforts was not a failure; instead, he explains to his nephew, it was merely bad luck, as he had convinced himself that the key lies in Gödel's First Incompleteness Theorem: Goldbach's Conjecture, he now believes, happens to be a rare instance of an undecidable statement. This is of course a work of fiction, yet in reality there are indeed a few instances of statements in mathematics too that have been shown to be undecidable in ZFC (in the sense of independent of ZFC). E.g., Shelah showed in 1974 that the Whitehead Problem, a problem in group theory, is independent of ZFC [76]. Another problem independent of ZFC is Wetzel's problem, a problem on the cardinality of a set of analytic functions fulfilling certain conditions: as shown by Erdős, it depends on the truth of the continuum hypothesis [1, pp. 132–134].[1] Another more recent example from 2011 is the proof by Farah [38] and Phillips and Weaver [71] that the existence of outer automorphisms of the Calkin algebra depends on set theoretic assumptions beyond ZFC: in particular, there exist outer automorphisms assuming the continuum hypothesis, while assuming Todorčević's Axiom all automorphisms are inner.

On a different note, within computability theory, undecidable problems in the sense of not effectively solvable, that is, computational problems for which there cannot exist a computer program that always gives the

---

[1] The proof was very recently formalised in Isabelle/HOL by Paulson [70].

correct "yes" or "no" answer, such as Hilbert's *Entscheidungsproblem* or the Halting Problem for Turing machines, are uncountably many, yet, once again, most such known statements are within logic rather than within mainstream mathematics. The group isomorphism problem is a notable example in combinatorial group theory, with more known examples in group theory, topology, linear algebra and analysis.

Such "anomalies" remain in the sidelines and it is generally accepted that foundations are not usually considered in mainstream mathematics research; this is normally inconsequential.

But nowadays foundations play another role, too: they are blueprints for proof assistants (also known as interactive theorem provers), which brings us to the main topic of this paper.

## 1.2 Proof assistants and mathematics

There are several reasons why formalising mathematics with a proof assistant can be useful. I have elaborated some reasons in an article [55]; I summarise these below. I would also like to point to a new preprint by Buzzard summarising his upcoming invited talk at the *2022 International Congress of Mathematicians* [17] as well as an earlier opinion piece by the same author for the *Notices of the London Mathematical Society* [14].

Obviously a first reason is verification. This however mainly applies to formalisation of research results, since otherwise the material has usually been checked by a great number of people over the years. A second reason is that contributing to the libraries of formal proofs amounts to the creation of a database with a huge potential. Formalised material could be used in the future to create new tools with the help of artificial intelligence to discover new mathematical results. A vision for the future is the creation of an interactive assistant that would provide "brainstorming" tips to research mathematicians in real time assisting them in the process of discovering (or inventing) a new result. In any case, a library written in code offers many possibilities: it is something we can modify, interact with, reuse, in contrast with a "physical" library consisting of printed books. A third reason is that the process of formalising in itself can help the user gain brand new insights even in already familiar topics. This is not only because a formalised proof must be written in a very high level of detail, but also because using new tools forces to look at familiar material from a new perspective. Last but not least, formalising mathematics with a proof assistant can also serve educational purposes.

An important milestone that indicates that formalised mathematics and proof assistants are gradually becoming integrated into mathematical practice is that the new 2020 Mathematics Subject Classification [3] includes for the first time [29] a new class (68Vxx) referring to topics such as computer assisted proofs, proofs employing automated/ interactive theorem provers, formalisation of mathematics in connection with theorem provers.

Some interesting very recent developments in the area of formalisation of mathematics with proof assistants are listed below. Summarising the most notable developments of the last couple of years in such a rapidly developing field would be no easy task; this is a collection of some highlights that I am aware of rather than a comprehensive review—I should apologise in advance for any omissions. The interested reader is also invited to explore e.g. the Archive of Formal Proofs and mathlib, the main databases for Isabelle and Lean respectively, that are growing at a very fast pace, day by day. Monthly progress in mathlib is summarised in the Lean Community Blog. Other extensive libraries are e.g. these of Coq and Agda.

A very important recent development is the *Liquid Tensor Experiment*: it involves the formalisation (in Lean) of material from the area of Condensed Mathematics. This is a theory that Clausen and Scholze started to develop almost four years ago: it claims that topological spaces are the wrong definition, and that they should be replaced with the different notion of condensed sets. Condensed abelian groups constitute a variant of topological abelian groups, but with more convenient properties. In late 2020, Scholze posed a challenge [74] which was realised by the Lean community very soon: in subsequent blogposts ([75] and more recently [19]) progress on the project has been reported. This achievement has gained extensive publicity and has been covered by *Nature* [18] and *Quanta* [48].

This is not the first important work by the Fields medalist Peter Scholze that was formalised in Lean: Perfectoid spaces, a special kind of adic spaces of fundamental significance introduced by Scholze in 2012 [73] have recently been formalised by Buzzard, Commelin and Massot [15].

There has been considerable recent activity in the area of number theory: *Lean Forward: Usable Computer-Checked Proofs and Computations for Number Theorists* led by Jasmin Blanchette is a major ongoing research project funded by the Dutch Research Council (NWO) that got launched in 2019 and will continue until 2023. The goal of Lean Forward is to collaborate with number theorists to formalise research-level theorems and to address the main usability issues that mathematicians are confronted with in their efforts to adopt proof assistants in their work. One of the major results of the project is the formalisation in Lean of Dedekind domains and class groups of global fields by Baanen, Dahmen, Narayanan, and Nuccio Mortarino Majno di Capriglio [5]. Another very notable development is the formalisation of the solution to the cap set problem by Dahmen, Hölzl, and Lewis [20] in Lean: this is a result by Ellenberg and Gijswijt published in the Annals of Mathematics in 2017 [35] addressing the size of subsets of fields that contain no 3-term arithmetic progressions.

Other recent, important progress in number theory involves the formalisation of a substantial amount of material in analytic number theory in Isabelle/HOL by Eberl [31].

Han and van Doorn have formalised in Lean the independence of the Continuum Hypothesis [45]. More recently, Gunther, Pagano, Sánchez Terraf, and Steinberg completed a formalisation of the above result in Isabelle/ZF [42].

Edmonds, Paulson, and the present author have recently formalised Szemerédi's Regularity Lemma, a major result in extremal graph theory [32]. We employed this to formalise the proofs of the Triangle Counting Lemma and the Triangle Removal Lemma and finally prove Roth's Theorem on Arithmetic Progressions, a major result in additive combinatorics on the existence of 3-term arithmetic progressions in subsets of natural numbers [33, 34]. Independently, and around the same time, Dillies and Mehta formalised the aforementioned results in Lean following a different approach [26]; their formalisations will be incorporated into mathlib in the near future.

An upcoming Special Issue on Interactive Theorem Proving in Mathematics Research of the journal *Experimental Mathematics* contains a number of papers on new contributions. Among these are a paper by Džamonja, Paulson and the present author [30] discussing formalisations [68, 69] of a number of research results in infinitary combinatorics and set theory (more specifically in ordinal partition relations, a field that deals with generalizations of Ramsey's theorem to transfinite ordinals) by Erdős and Milner [36], Nash-Williams [63], Specker and Larson [60], leading to Larson's proof of an unpublished result by Milner asserting that for all $m \in \mathbb{N}$, $\omega^\omega \to (\omega^\omega, m)$. This is of interest not only because it involves formalisation of material within an area never formalised before (to our knowledge), but also because it is a demonstration of working with Zermelo-Fraenkel set theory in higher-order logic [67], as all the formalisations were done in Isabelle/HOL. Another paper in the issue, by Li, Paulson, and the present author [59], explores the formalisation of relatively modern mainstream research papers in number theory and analysis discussing our formalisations [56, 57, 58] in Isabelle/HOL of certain irrationality and transcendence criteria for infinite series from three different research papers (by Erdős and Straus [37], Hančl [46], and Hančl and Rucki [47]). A very important work on a different topic, also included in the aforementioned special issue, discusses the formalisation in Lean of Grothendieck's schemes in algebraic geometry by Buzzard, Hughes, Lau, Livingston, Fernández Mir, and Morrison [16]. A short time later, schemes got formalised in Isabelle/HOL, too, by Bordg, Paulson, and Li [12, 13]: to make up for the lack of dependent type theory in Isabelle/HOL, locales were employed instead to achieve the required level of expressiveness.

## 1.3 Proof assistants and formal systems: back to the main question

Like a driver who ignores the details of how their car engine works but still drives safely to their destination, the typical mathematician proof assistant user may not always be concerned with all the ramifications of the formal system their proof assistant of choice is based on. This is typically not true of proof assistant developers. In any case, the issue of foundations, usually hidden in the background, becomes inevitably more relevant when using a proof assistant compared to when doing mathematics with pen and paper.

Today there is a number of different proof assistants, based on different formal systems, that provide important libraries of formalised mathematical proofs. We can distinguish three big families: the proof assistants based on set theory (e.g. Mizar, Metamath); these based on simple type theory (e.g. HOL4, HOL Light, Isabelle); and these based on dependent type theory (e.g. Coq, Agda, Lean, PVS). The interested user who would like to see a direct comparison between their corresponding languages through examples of formalised material is referred to *Formalizing 100 Theorems*, a list of central mathematical theorems formalised in different proof assistants, maintained by Freek Wiedijk on his website. (Another source for comparisons, [81], though very useful, being one and a half decades old it is dated, as Lean [62], a very widely used proof assistant especially among mathematicians, would enter the picture seven years later, in 2013).

This pluralism invites us to explore to what extent logical foundations might have an effect here, i.e., within which aspects of proof assistants they manifest themselves (if at all). Thus, inevitably we are returning to the original question of the title: the question of *how formal systems themselves matter in practice for proof assistants and what they can do for mathematics*, possibly transforming the contemporary and future mathematical landscape via proof assistants.

Such a transformation may refer to the kind of research work we do; most likely it will relate to various different aspects of mathematical practice, that is, not only (or not necessarily) *what* we do but also *how* we do it. It may, e.g., refer to an evolution of the reviewing system by having the authors submit code containing formalised versions of their results along with their proofs written in "usual" mathematical language (Hales's massive Flyspeck project that succeeded in 2014 [44] to formally verify using HOL Light and Isabelle/HOL his 1998 proof of the Kepler conjecture [43] is of course well-known; we can mention a couple of examples in recent pure mathematics research, still rather isolated thus pioneering, where the formal proofs were submitted simultaneously with the original result, such as the work [50] by Kjos-Hanssen, Niraula and Yoon in metric geometry with the result formalised in Lean and the work [40] by Gouëzel and Shchur in Gromov-hyperbolic spaces with the result formalised in Isabelle/HOL). The possible upcoming transformation may moreover refer to teaching, learning and dissemination approaches in mathematics; to social aspects of large collaborative projects; or even to an increase of interest in the foundations of mathematics among mathematicians.

But let us listen to what a few of the protagonists have to say—in their own words.

## 2 Interviews

**Jeremy Avigad, Department of Philosophy and Department of Mathematical Sciences at Carnegie Mellon University & Charles C. Hoskinson Center for Formal Mathematics.**

**K.-A.** The new Charles C. Hoskinson Center for Formal Mathematics at Carnegie Mellon University of which you are the director, was inaugurated in September 2021. The center aims at the advancement of mathematical research by facilitating access to knowledge and resources. To this end, one of the main goals of the center is to support the development of Lean's library of formalised mathematics and of new tools to help convert mathematical statements from natural language to a formal language, as well as the creation of educational resources for the dissemination of these tools. Would you like to elaborate on how this will contribute to the evolution of mathematical practice (and the practice of related disciplines) as we know it?

**Avigad.** It is important to recognise that formalisation is not just about checking correctness. Building a digital mathematical library is a wonderfully collaborative activity, and the result is an important communal resource: a permanent, precise repository of mathematical knowledge, interlinked, searchable, and surveyable at any level of detail. It's like building the library of Alexandria and making it fireproof by putting it in the cloud. The resources—not just the definitions and theorems but also the algorithms and software tools that act on them—are freely available to anyone with an internet connection. They can be used to support education and discovery as much as verification.

The mission of the Hoskinson Center is to help make that technology accessible to as wide an audience as possible. Infrastructure and library development are an important part of that, but our main focus is on education and dissemination.

**K.-A.** Based on your rich experience with various different proof assistants (Isabelle/HOL, Coq and, more recently, Lean) what are, in a nutshell, the strengths and weaknesses of each system? What are some improvements that you would like to see in future versions of Lean?

**Avigad.** I don't like to make comparisons. Proof assistants are like programming languages. Everyone has their favourites, and people spend far too much time fighting over which ones are better. All three systems you mention are wonderful. All of them have yielded fundamental, groundbreaking contributions to our understanding of formal methods and what they can do.

In recent years, my focus has been on Lean. It is a beautifully designed system and it has attracted a number of energetic, enthusiastic young mathematicians and computer scientists. My favourite thing about Lean is the community that has grown around it.

In general, though, proof assistants are still too hard to use. Using the technology requires too much dedication; it's impossible to just pick it up for casual use. We need better libraries, interfaces, search engines, automation, and educational resources. Progress has been slow but steady. We'll get there.

**Jasmin Blanchette, Department of Computer Science, Vrije Universiteit Amsterdam, VeriDis group at Loria, Nancy, Max-Planck-Institut für Informatik, Saarbrücken, & Institut für Systemsicherheit, Universität der Bundeswehr München.**

**K.-A.** A benefit of using simple types instead of dependent types is the more practical implementation of efficient automation. Isabelle uses simple types and Isabelle/HOL, encoding higher-order logic, is implemented with Sledgehammer's [7, 11] automation, that calls several external automated theorem provers, giving Isabelle/HOL an advantage over other proof assistants. However, Isabelle/ZF, encoding Zermelo-Fraenkel set theory, does not feature Sledgehammer. As a leading expert in automated theorem proving and in particular Sledgehammer, would you like to elaborate on the obstacles (if any) to the implementation of Sledgehammer to Isabelle/ZF?

**Blanchette.** To adapt Sledgehammer to Isabelle/ZF, the three main modules of the existing Sledgehammer would have to be revisited:

(1) the relevance filter, which selects (typically) a few hundred facts (lemmas, definitions, ...) from the (typically) thousands available in background libraries;

(2) the problem translation module, which encodes Isabelle formulas into the logic of the target automatic theorem prover (ATP);

(3) the proof reconstruction module, which takes an ATP-generated proof and translates it into an Isabelle proof.

The relevance filter would be straightforward to port to ZF. It would need to be told which symbols are ZF primitives, so that it ignores them when computing the relevance of a fact with respect to the formula to prove.

The problem translation module would need the most adaptation, because it would need to translate from a different logic. I know Cezary Kaliszyk had a prototypical translation running, and the MizAR "hammer" for the Mizar proof assistant also performs a translation for another set theory, so this is possible. Most automatic theorem provers support first-order logic, and set theory is formulated in terms of that logic.

Proof reconstruction depends on having strong built-in automation directly in Isabelle/ZF that can reconstruct arbitrary ATP steps. For Isabelle/HOL, we integrated the Metis prover by Joe Hurd. One option would be to integrate Metis in Isabelle/ZF in much the same way.

Barring a lot of engineering, I see nothing really standing in the way of a Sledgehammer for Isabelle/ZF.

### Frédéric Blanqui, INRIA, Laboratoire Méthodes Formelles, Université Paris-Saclay.

**K.-A.** You are the main proposer and Chair of the European COST Action CA20111 EuroProofNet (European Research Network on Formal Proofs) that got launched recently, in October 2021, and has over 200 participants from 30 countries. One of the main objectives of the Action is to work towards the ambitious task of achieving interoperability between different proof systems (like Coq, Isabelle/HOL, HOL Light, Agda, Lean, Matita and others). To this end, a new common logical framework (Dedukti) will be developed, along with new tools for inter-translation of proofs formalised in various different systems to and from the new common logical framework. What are some of the required characteristics of the new formal system? What do you expect will be some of the most challenging obstacles to overcome so as to achieve inter-translation between the new "global" formal system and other systems?

**Blanqui.** Various logical frameworks have already been developed in the past. The most prominent one, adopted by all mathematicians now, is first-order logic. In first-order logic, each mathematical theory is characterized by some propositions called axioms. E.g., Euclidean geometry and hyperbolic geometry can both be expressed in first-order logic using different axioms.

However, first-order logic suffers from a number of weaknesses. E.g., it is uneasy to express properties and proofs of objects with binding constructions, or modern categorical constructions. That is why, nowadays, many proof assistants are not based on first-order logic but rather on higher-order logic or dependent type theory, that is, a type theory where types can depend on objects.

Similarly, Dedukti is based on the simplest dependent type theory possible. But, in Dedukti, in contrast with other systems, typing is done modulo user-defined equations. Hence, while in first-order logic, theories are characterized by the symbols they use and their axioms, in Dedukti, theories are characterized by the symbols they use and their equations.

Reasoning modulo equations on types is the key feature that allows one to express in Dedukti the proofs of many systems: first-order logic and its theories, higher-order logic, as well as complex type systems like Agda and Coq.

However, we have to express in the same way the features (polymorphism, predicate subtyping, proof irrelevance, impredicativity, etc.) that are common to all systems, so that we can more easily detect the proofs that can be translated from one system to the other. E.g., Euclidean geometry and hyperbolic geometry only differ by one axiom. Hence, a proof in Euclidean geometry not using Euclid's 5th axiom can be readily translated into hyperbolic geometry.

There are many theoretical and practical challenges to scale up and fully handle all current proof assistants. A first challenge is to extend the set of proof assistant features that can be expressed in Dedukti. A first step is the article "Some axioms for mathematics" at the 2021 FSCD conference [9]. Another very important challenge is to make the translated proofs usable. This in particular requires to align the concepts used in the source system with the ones used in the target system.

## Kevin Buzzard, Department of Mathematics, Imperial College London.

**K.-A.** Continuing a celebrated career in algebraic number theory, in recent years you have also been enthusiastically working on formal proof verification—an area that used to traditionally attract mostly computer scientists rather than mathematicians—using Lean. You are very actively involved in teaching mathematics undergraduates how to use Lean to formalise proofs, and you have been supervising many student projects in this direction, too; a new generation of mathematicians, in their formative years, is becoming accustomed to formalisation and interactive theorem proving. Watching your students learn, how do you think using a proof assistant affects the development of their mathematical thinking? Have you noticed that it may increase the students' attention to detail, or spark an interest in foundational questions? In your experience, is getting accustomed to Lean's dependent type theory actively affecting how students think about mathematics?

**Buzzard.** Here is what I think is going on: Students learn very quickly when coming to university that there are things which are acceptable to say in some courses, but not acceptable in others. E.g., the fact that the derivative of sine is cosine would be a perfectly acceptable claim in a 1st year mechanics course, but would probably need a detailed proof in a 1st year analysis course. I teach students how to formalise mathematics, but I am not convinced that it changes the way they solve mechanics questions one jot! I am not even convinced that it changes the way they work in their other pure mathematics courses. Perhaps it changes things slightly: maybe they are more careful with edge cases, maybe they factor out sublemmas a bit more, maybe they write more constructively. But what does this buy them in practice? Mathematicians are sometimes sloppy with edge cases—e.g., in Atiyah and MacDonald's book on commutative algebra, in their proof of the Artin-Tate lemma they do an induction where the inductive step works fine but the base case does not! An undergraduate pointed this out to me when they were formalising it in Lean and I was shocked, but Lean was right. But of course edge cases to mathematicians are just boring and trivial. Mathematicians don't need to compile their work so who cares if it is written in a modular manner. And of course, most mathematicians have no idea what constructivism is. I wonder therefore whether all that is happening is that undergraduates are learning a third "mode"—there is "applied mathematics mode", "pure mathematics mode" and then "formalisation mode". One thing I've learnt from undergraduates (showing me their problem sheets) is that pure mathematics, even at undergraduate level, is sometimes extremely difficult to formalise. A student might observe that a question on an algebraic topology example sheet which just comprises of a couple of pictures would be extremely hard to formalise, e.g., it might even be extremely hard to formally show that the question is well-defined; I see a picture of a torus being butchered in some way, but how do we know that the resulting object is truly independent of the details of the butchering? However, the students know what the lecturer is looking for in a solution, because they have seen how the lecturer does analogous questions and they understand that the point is that they are to copy the techniques and this will be an acceptable argument in this context. In particular, the student knows not to be in "formalising mode" when solving some of the problems, they just switch back into "maths how it is done in class in practice" mode.

Another observation would be that there is this inconvenience of type theory: types cannot intersect nontrivially. Sets of course can, so in ZFC, it's easy to switch between the notion of a "thing" being both a subgroup and a group at the same time. In type theory this is not so easy. Students learning Lean are learning how to do mathematics in type theory, but in practice a mathematician when working on paper just uses whatever theory is most convenient for the situation at the time; we are under no pressure to be consistent in practice.

I think that in conclusion I would say that I am teaching the students how to think carefully and pedantically, which is definitely a useful skill, but that mathematics is not only about this.

## Johan Commelin, Mathematisches Institut, Albert-Ludwigs-Universität Freiburg.

**K.-A.** In parallel with your research in algebraic geometry and algebraic number theory, you are very actively involved in the Lean theorem prover community. You are currently leading the Liquid Tensor Experiment, having taken up a challenge posed by Peter Scholze, formalising cutting-edge research mathematics. In your experience so far, how do think Lean's dependent type theory, as an underlying logical formalism, has been helpful in formalising this area of mathematics in terms of expressiveness? E.g., overall, what have you and your collaborators found more challenging: expressing definitions in the language of Lean or actually working out proof steps? What do you think are some of the pros and cons of Lean's dependent type theory and what are some changes that you would like to see in future versions of Lean?

**Commelin.** The Liquid Tensor Experiment showed that within a reasonable amount of time, difficult state-of-the-art mathematics can be formally verified; providing a tangible benefit for the working mathematician.

In my experience, this is the first time that I was genuinely *assisted* by the computer in understanding a proof. Attempts to penetrate the proof using pen and paper went nowhere. But by gradually building the formalisation (with the help of a dozen other people) I gained a solid understanding of all the moving parts, and how they precisely fit together. This proof is extremely complicated, and walks a very fine line. In that sense, it comes as no surprise that Scholze had not received any feedback on this proof from within the mathematical community more than a year after publishing this proof on his website. At the same time, Scholze explains that he had some small lingering doubts about this proof, precisely because of this complexity.

The formalisation process, and all the interaction that ensued between Scholze and the formalisation team also led to a better understanding of the structure of the proof. Scholze wrote in a blogpost [75] that for him the arithmetic nature of the proof of this analytic result was clarified. Besides that, a technical ingredient was significantly simplified.

I do not know enough about type theories to give a meaningful statement about how important dependent type theory is for the Liquid Tensor Experiment. I recognize the following important factors in the success:

(1) The community. When Scholze posed his challenge (in all proof assistants), there was an enthusiastic response in the Lean community.

There were sufficiently many mathematicians with a good working-knowledge of Lean, which led to fast and fruitful collaboration.

(2) Lean's mathematical library, mathlib. This is a large and coherent library of many topics in undergraduate mathematics, which means that all these developments can be imported and used simultaneously. Both because it is developed in a single repository via a coordinated pull request system, and because the mathlib continuous integration process ensures no name collisions and no bad interactions between simplifier lemmas.

We used extensively the sections on algebra, topology, analysis, category theory, and homological algebra. Futhermore, when things did not work right, we could refactor mathlib and improve the library. Algebra (monoids and groups), analysis (semi-normed groups) and homological algebra (complexes) were refactored as part of the project. These refactors would be merged upstream in an efficient process, which was crucial for keeping the ball rolling.

The success of the Liquid Tensor Experiment is very much a success of the community around Lean and mathlib.

## Manuel Eberl, Computational Logic Group, University of Innsbruck.

**K.-A.** As a very experienced user of Isabelle and a prolific contributor to the Isabelle Libraries and the Archive of Formal Proofs, how useful has Sledgehammer been in your everyday formalisation work?

**Eberl.** I do use Sledgehammer a lot, albeit less so than other users. When I do use it, it is mostly as a tool to find relevant facts to my current goal. When I think the current goal should be doable by the automation given the correct rules but I'm not quite sure what the correct rules are, I call Sledgehammer, and if it finds a proof, I then inspect the facts it uses and construct a "proper" proof from them. I tend to prefer that to the kind of proofs that Sledgehammer generates. Other people tend to use Sledgehammer regularly to find proofs and keep them that way. So I think I would not miss it quite as much as a lot of other people if it did not exist, but I certainly use it often enough that it would impede my productivity if it did not exist.

**K.-A.** Is the lack of Sledgehammer the reason that Isabelle/ZF is much less used (compared to Isabelle/HOL), or are there other reasons related to the expressiveness of Isabelle/ZF?

**Eberl.** First of all, I must say that I have never used Isabelle/ZF, so take everything I say with a huge grain of salt. My impression is that there is no problem at all with the expressiveness of Isabelle/ZF. The major reason why it is not used is probably simply that the libraries and tooling (proof automation, probably definitional tools) are nowhere near as nice as those of Isabelle/HOL. As for why that is, I suspect that a large amount of it is simply historical: Isabelle/HOL got quite popular, so many people flocked to it and built libraries and tools, and that made it even more popular etc. Nothing comparable happened for Isabelle/ZF.

From what I understand, there are technical issues with using "untyped" logics such as ZF in a theorem prover. Working with the "naked" logic is possible, but tedious. For productive work, you want something like Mizar's soft types, and overloaded operators, and management of algebraic structures like groups (perhaps using some analogue of type classes). All of that takes quite a bit of engineering. I understand that there were (or still are) people working on such things, and I'm sure they can in principle be brought into a state were they work well enough to make working in Isabelle/ZF just as pleasant as in Isabelle/HOL, but it is a lot of work, and it takes quite a lot of energy to get it to a "self-sustaining" point where it is good enough that it will attract other people to work on it.

**K.-A.** Do you think you can make a prognosis about the possible limitations of simple type theory?

**Eberl.** In the kind of mathematics that I formalise, I have not really encountered any hard obstacles due to using simple type theory. There are definitely hard limitations, such as the inability to quantify over types, which makes some definitions basically impossible to write down. And as far as I know, there are also other logical issues when it comes to formalising category theory or advanced set-theoretic arguments like forcing, but I do not know much about these fields. In my experience, such things do not crop up too often in practice. The more interesting limitations are where something is possible to do in simple type theory, but becomes significantly more painful.

The one part of the Isabelle library where such issues become most apparent to me is in abstract algebra: the obvious way to define, e.g., a group is as a type with a binary operation and a neutral element that fulfil certain laws. The type classes in Isabelle/HOL make this approach very comfortable, which is why that is the way it is done in the standard library. But as soon as we want to talk about different groups and their relations to one another and their subgroups etc., this does not work without dependent types. Instead, one can define groups with an explicit carrier set and carry the group operations and the group axioms around manually. This is the approach that the HOL-Algebra library (also in Isabelle/HOL) follows, but it is much less pleasant to use than the type-based approach from the standard library.

A similar situation exists, e.g., for topologies: there is a type-class-based version where the topology comes from a type class (where you can just write "open $X$" for "the set $X$ is open"), and a more general one where a topology is a separate object (where you write "openin top $X$" for "the set $X$ is open with respect to the topology top"). Having two separate libraries that do the same thing in different ways is always problematic for maintainability and usability. There is some tooling that allows transferring results between the two approaches, but it is unfortunately not as pleasant to use as one would want.

In any case, these problems are not show stoppers. They may force you to write down some things in a less natural fashion than you would have to in a stronger logic, but this extra effort always has to be seen in conjunction with the advantages of simple type theory: as far as I am aware, simple type theory-based systems like Isabelle/HOL tend to have better performance and better proof automation than systems using stronger logics. Depending on the concrete application, this may offset the less expressive logic.

## Sir Timothy Gowers, Collège de France, Department of Pure Mathematics and Mathematical Statistics, University of Cambridge, & Trinity College, Cambridge.

**K.-A.** You are a strong supporter of proof assistants and the formalisation of mathematics; for many years, during your celebrated career in mathematics, you have been advocating the vision of developing an interactive tool that would provide direct assistance to working mathematicians in their everyday research by offering ideas, hints, information related to the work at hand and proving auxiliary results [39, 41, 4]. Given the state of the art of the tools and the most recent advances in the area of proof assistants and formalisation, how optimistic are you that this vision is getting closer to realisation?

**Gowers.** As background, I should say that my main interest in this sphere is not so much formalisation as fully automatic proof discovery, but the two are sufficiently closely related that I am certainly interested in following what other people are doing in the formalisation area.

My answer to your first question is that I am very uncertain. I think there is certainly the potential for a lot of non-revolutionary progress, which I think might be enough to develop tools of genuine use to mathematicians. Indeed, I hope to contribute to that development by continuing to work on human-oriented theorem proving, but of course there are other ways that progress may well be made, such as machine learning. (My hunch about machine learning is that there is a fundamental limitation to what it can do if it is operating entirely on its own—roughly speaking corresponding to the need to have non-obvious mathematical ideas. So I expect to see impressive progress in the short term but also to see a plateau being reached unless it is combined with more traditional approaches. But it's always dangerous to make predictions like that.)

My short answer to your question is that I am optimistic that there will be exciting progress over the next ten years or so, but I think we are probably two or three big breakthroughs away from a system that can solve interesting research problems. (But a lot depends on what one means by "solve" here, and in particular on how much human assistance there is.)

**K.-A.** What are some theorems or research areas that you would like to see formalised in the near future?

**Gowers.** I suppose I'd answer that I don't have a direct interest in any particular part of mathematics becoming formalised, because the areas I work in tend to be ones where we are not too worried about whether the main results are correct. (I contrast that with some areas where major results sometimes depend on other major results that have never been properly written down, and things like that. Such situations are quite rare in combinatorics.) So I could go in one of two ways. Either I'd say that for the health of mathematics it would be good to identify the more "troubled" areas where the foundations are potentially shaky and put those right, or I'd simply say that I'd find it personally satisfying to see the main results in my own area formalised. Or I could be more selfish still and say that in the past I have written some pretty complicated papers that almost certainly have details that aren't fully correct, so it would be a good feeling if some of them were formalised—if I ever found out that somebody was interested in doing that, I would be more than happy to cooperate in finding any necessary corrections. (I do not believe that any of my papers are incorrect in a worrying way, but some of the fixes could be quite hard to find.)

**K.-A.** You initiated the first Polymath project in 2009. Do you think that thanks to the rise of proof assistants massively collaborative mathematics may soon become customary?

**Gowers.** I think that what's going on with formalisation could be described already as a massively collaborative project, though it's not precisely the same as mathematics. The way things look at the moment, people are too wedded to the current reward structures of mathematics for the massively collaborative approach to become the main way of operating, and it can be hard to organize a successful project unless one has a strong online presence. So I think that there will continue to be Polymath projects, but I don't think they are going to become mainstream in the near future.

**Peter Koepke, Mathematisches Institut, Rheinische Friedrich-Wilhelms-Universität Bonn.**

**K.-A.** Proof assistants use formal languages that are reminiscent more of computer code rather than "natural" mathematical language and notation that is familiar to mathematicians. The goal of natural proof assistants is to remedy this issue, thus making formalised proofs more easily readable and proof assistants more user-friendly to mathematicians. An effective combination of the formal rigour of logical calculi behind proof assistants with the use of natural mathematical language and notation as in usual mathematical textbooks and research papers would be an extremely powerful development. As one of the creators of the Isabelle/Naproche [24, 25] natural proof assistant, would you like to share some insights about choosing an appropriate underlying logical formalism that can nicely combine with natural language?

**Koepke.** We focus on the natural language and structuring of mathematical texts as they are really employed in mathematical publications. In an appropriate context of definitions, notations, and axioms, Naproche is, e.g., able to directly accept, process and proofcheck (a LATEX source of) the following formulation of the open mapping theorem from complex analysis, including natural language phrases and mathematical symbolisms:

**Theorem 2.1** (Open Mapping Theorem). Assume $f$ is a holomorphic function and $B_\varepsilon(z)$ is a subset of the domain of $f$. If $f$ is not constant on $B_\varepsilon(z)$ then $f[B_\varepsilon(z)]$ is open.

This kind of mathematics is usually modelled by first-order logic and some version of set theory. We have inherited the first-order approach from our predecessor system SAD (System for Automated Deduction, by Varchinine, Lyaletski, and Paskevich [79]). An important advantage is that the leading automated theorem provers which Naproche continuously calls for minor proof tasks use first-order logic themselves. As natural language and the language of mathematics use "soft types" it may be better in the long run to switch to an appropriate formal language between first-order logic and type theory.

**K.-A.** What are the reasons behind your choice of Isabelle?

**Koepke.** Originally we have chosen Isabelle as a proof development environment in which we can comfortably run and use the Naproche program. So far there has been no connection to the typical Isabelle logics like Isabelle/HOL. We are, however, collaborating with Makarius Wenzel to find ways to map logical entities of Naproche to Isabelle and vice versa. We have an experimental setup where Naproche is invoking the Isabelle/Sledgehammer mechanism to discharge certain proof obligations.

**K.-A.** How has your long career in set theory and foundations of mathematics influenced your taste in logical formalisms for proof assistants?

**Koepke.** My experience with first-order set theory of course came in handy for dealing with SAD/Naproche formalisations and with the logical mechanisms of the system. Similar to Naproche, the working language of set theory introduces all sorts of notions like numbers, functions, structures, etc. on top of the frugal language of the axioms. But if other formalisms accommodate actual natural language better I shall happily adopt them. The trouble with natural (mathematical) language are the freedoms and ambiguities that people use to convincingly express their intuitions and arguments. First-order logic is able to deal with all sorts of exceptions by "if ... then ... otherwise ..." constructs, whereas other formalisms might force users to write in a more regulated and possibly "unnatural" language.

## Assia Mahboubi, INRIA, Gallinette Team, Nantes & Vrije Universiteit Amsterdam.

**K.-A.** You are a very experienced user of the Coq proof assistant and you are particularly interested in the interplay between computer algebra systems and formal proofs. Your new EU-funded FRESCO project ("Fast and Reliable Symbolic Computation"), that started in November 2021, aims to explore whether computer algebra systems could be both reliable and fast. To this end, the programming features of proof assistants will be enriched to become efficiently combined with computer algebra systems, while compatibility with logical foundations will be preserved. Undoubtedly, the combination of the computational power of computer algebra systems with the formal reasoning tools provided by proof assistants based on various logical formalisms would be a very powerful combination. Would you like to give a few examples of applications in mathematical practice and research where we could see significant advances thanks to this interplay? And would you like to comment on which logical formalism(s)/formal system(s) you believe would be more preferable or practical to combine with computer algebra systems?

**Mahboubi.** From experimentation to proofs, there is a tremendous momentum right now for computer-aided mathematics. The use of computers for formulating conjectures, but also for substantiating proof steps, pervades mathematics, even in its most abstract fields. Most of these computer proofs are produced by symbolic computations, using computer algebra systems. Sadly, these systems suffer from severe, intrinsic flaws, key to their amazing efficiency, but preventing any flavour of post-hoc verification.

The field of number theory is an emblematic example of this change of era, and computers have become a crucial instrument for basic research in number theory, both for designing conjectures and for proving them. The Birch and Swinnerton-Dyer's conjecture about the rational points of an abelian variety, is a notorious example of a computer-shaped hunch, based on (at the time) intensive calculations. This fertile conjecture has sparked fundamental contributions, although if the problem remains open at the time of writing: the Clay Mathematics Institute even advertises a 1 million dollar bounty for its solution. Regarding published proofs, prominent examples include Bhargava's work in the sub-field of number theory called "geometry of numbers", crowned by a Fields medal in 2014 or Helfgott's groundbreaking proof of the long-standing Ternary Goldbach conjecture, in analytic number theory.

Verified (and fast) computer algebra would of course provide a principled way to assess the correctness of such achievements, when the social process of peer-reviewing just falls short of evaluating the proofs produced by computers. But it would also provide an invaluable tool for gaining confidence in the intuitions shaped by computer calculations. These calculations can be of a very diverse nature: evaluation of formulas, simplification of algebraic expressions, plots, etc. But these can all go very wrong, and for subtle reasons, which pertain to the semantic of symbolic computations (or to the lack thereof) rather than to the bugs in the usual sense, that of programming errors. Another important application I can foresee is the validation of handbooks and databases, gathering collections of mathematical objects together with their properties, like the celebrated NIST Handbook of Special Functions or the reference $L$-functions and Modular Forms Database, in number theory.

Modern verification tools can be classified according to the expressivity of the logical language available to state specifications, which in turn impacts their ability to automate proof search. E.g., variants of propositional logic have earned their stripes in hardware design, but specifying and verifying cyber-physical systems requires first-order logic and beyond. Verifying computer algebra in the large is more demanding, as elementary specifications will casually involve quantifying over objects such as "finite fields of an arbitrary characteristic $p$", with a formal integer $p$. Such a parametrisation is typically beyond the skills of computer algebra systems; they only provide concrete instances of these fields, for concrete values of $p$, as this prime integer controls algorithmic choices in modular arithmetic. In fact, verifying computer algebra in the large calls for a first-class, representation of hierarchies of mathematical structures in the logic, a feature which is available in dependent type theory.

**Ursula Martin, School of Informatics, University of Edinburgh & Wadham College, Oxford.**

**K.-A.** Your celebrated career covers many areas in theoretical computer science and formal methods; in recent years, supported by an EPSRC Fellowship, you have also been investigating the nature of mathematical practice as a social machine with your project "The Social Machine of Mathematics". In addition to various online databases (such as arXiv, Google Scholar, ResearchGate, MathSciNet, Orcid...) having revolutionised access to information and knowledge, various online platforms such as MathOverflow, MathStackExchange, GitHub and Zulip (and sometimes, even social media) actively facilitate communication and collaboration. The latter tools have been broadly adopted by the proof assistant community and contributed to its activity and success. Do you think that mathematical practice is, in this way, being radically transformed? If yes, do you perhaps see more working mathematicians taking more interest in the foundations of mathematics due to interest in proof assistants? And do you think that the existence of many different proof assistants is affecting (either accelerating or hindering) the progress of this transformation?

**Martin.** It's hard to recall, even for those of us old enough to remember, what the practice of mathematics was like before the personal computer; before email; before the internet; before LaTeX; before the arXiv; before collaboration sites like MathOverflow and GitHub; and before the massive expansion of mathematical publication that these now ubiquitous tools have enabled, with mathematical publications quadrupling since 1996.[2]

Yet what has remained remarkably stable is the notion of an academic mathematical paper: a document, now digital rather than paper, but otherwise still published in a journal sponsored by a recognised entity, attributed to a small number of named authors, accredited by the journal's refereeing process, with a fairly standard kind of structure, content, argument and citation practice established within particular subdisciplines. While there are undoubted triumphs of the use of proof assistants—e.g., Gonthier's work on the Odd-Order Theorem, or Hales's proof of the Kepler conjecture—the contribution to published mathematical papers of proof assistants is at a low level, by contrast with the contribution of software such as GAP, Mathematica or MATLAB. Thus, while one might argue that academic mathematical practice has the potential to be radically transformed by proof assistants, it is less easy to claim that this has so far happened.

To look at the reasons for this, one might look, not at the nuances of particular proof assistants, but more generally at how technological, or

---

[2]Cf., e.g., the World Report on the Scimago Journal & Country Rank webpage, listing 53,564 mathematics publications in 1996 and 183,582 mathematics publications in 2019.

other, change comes about within the institutions involved in the creation of academic mathematics: universities, funding bodies, and publishers, and at how institutions, or individuals within them, are incentivized to bring about far-reaching changes to current practices, when such practices, e.g., journal publications, are so entrenched in traditional mechanisms for credit and reward. It is noteworthy, e.g., that Gonthier did his work while employed by a company (Microsoft Research), and one might speculate on whether Hales would have felt able to do his lengthy machine proof of the Kepler conjecture if he had not had tenure. By contrast, in commercial software development, if proof technology is needed to achieve the commercial ends of the company, the infrastructure and workflow can be restructured to accommodate it—cf., e.g., Facebook's use of the Infer static analyser, which is based on separation logic [27].

As to the impact of having many different software tools—proof assistants among them—certainly uptake and use of such tools might be enhanced if a single proof assistant became as ubiquitous and as taken for granted as LaTeX, Maple or Jupyter Notebooks, or indeed embedded in such platforms. Is the reason for a diversity of tools a positive one—the need the research community feels, at this stage of development, for diversity and for experiment rather than being forced into a common approach? Or is it rather a lack of institutional or individual incentives, enthusiasm, leadership or skill for the trade-offs and political activity involved in uniting behind a common approach?

## Lawrence C. Paulson, Computer Laboratory, University of Cambridge & Clare College, Cambridge.

**K.-A.** You are the Principal Investigator of the ERC Project "ALEXANDRIA—Large Scale Formal Proof For the Working Mathematician" which started in Autumn 2017 at the University of Cambridge and aims at the investigation of the formalisation of mathematics in practice using Isabelle/HOL and, more broadly, at contributing to the creation of a proof development environment attractive to working mathematicians. What do you think are the main takeaways from the project so far?

**Paulson.** ALEXANDRIA has made it possible to engage with the mathematical community and fully explore the issues surrounding the formalisation of mathematics. This engagement was primarily but not exclusively with Isabelle/HOL, since some experiments were also done using Lean. Roughly speaking, our objectives were to explore what could be accomplished using our tools, and where weaknesses were identified, to try to mitigate them.

I've been immensely satisfied with our progress. A group of people from quite different backgrounds have pursued a variety of subprojects, some individually and some in smaller groups, sometimes with outside collaborators. We accomplished things that some observers thought were impossible, notably the formalisation of schemes in Isabelle/HOL [12, 13]. Time and again we saw that advanced material such as Szemerédi's Regularity Lemma [32] weren't really that difficult and much of the difficulty we did encounter lay in the errors and inconsistencies of our source material.

A separate but essential effort was aimed at reducing the labour that goes into formalisation. This produced the SErAPIS search engine [77, 78], which is the first example of true intelligent search for an interactive theorem prover: material from all the libraries can be sought on the basis of concepts as well as syntax. Other research, less well advanced but deeply important, is aimed at using machine learning and other technologies to recognise proof patterns and assist the interactive process of formalisation by suggesting solution fragments.

**K.-A.** As one of the creators [65, 64] (and a very experienced user) of Isabelle, you have stressed many times that, when it comes to different formal systems, pluralism is both important and inevitable. At the same time, the answer to the question of whether simple [66] or dependent type theory is more practical for formalising mathematics (or rather: each area of mathematics) and what is the best way to use each formalism, is far from being clear. Would you like to elaborate a bit on this topic?

**Paulson.** I'm glad you say that "the question of whether simple or dependent type theory is more practical for formalising mathematics... is far from being clear". Plenty of people have made up their minds already. But that's premature in such a rapidly developing field. We have seen that we can go a long way in simple type theory (interestingly, with rather little reliance on axiomatic type classes). We have not encountered any no-go areas so far. We have the benefit of extensional functions and ordinary equality as well as strong automation, and full set theory when we need it. Dependent type theories are lacking in those particular areas, but have the benefit of greater expressiveness; and as they continue to evolve, who can say what things will look like in five or ten years?

## 3 A final comment

Our discussion has merely scratched the surface of this vast topic, yet it is already obvious that foundations, as the basis of proof assistants, can potentially be of substantial service to mathematical practice in many different ways.

But let us conclude with acknowledging that when it comes to computers and the future of mathematics, proof assistants and foundations are only one side of the story.

This is not only because computer algebra systems, natural language processing and automatic proof discovery can provide indispensable tools too, as already mentioned in the above discussions. This is because it appears that the axiomatic/ symbolic approach as conceived by Leibniz, following his vision of *calculus ratiocinator*, could take us only so far—even assuming we eventually manage to formalise "all" mathematics as wished in the QED Manifesto from the 1990es [2]. Progress seems to require the combination of alternative approaches. An interesting analogy due to Georg Gottlob is seeing "rule knowledge and logical reasoning versus machine learning, e.g., neural networks" as "left part of the brain versus right part of the brain": they have different, but complementary functions, the former inducing rationality and the latter inducing imagination and creativity.

To achieve the creation of interactive tools that would actively and efficiently help research mathematicians in their creative work, it is also new advances in artificial intelligence and machine learning that can promise novel developments in mathematical practice through their applications to automated theorem proving and proof assistants. E.g., pattern recognition tools from machine learning could find applications not only in searching the libraries of formal proofs, but also in recognising proof patterns and providing proof recommendation methods thus enhancing automation. Some promising efforts in this direction are, e.g., [8, 54, 61, 6, 72]. A new book by Holden [49] provides a comprehensive review of research applications on incorporating machine learning into automated theorem proving which has the potential of developing tools that could improve automation tools featured by interactive theorem provers.

The communities of machine learning and formal verification have indeed been growing increasingly close during the past few years. Some highlights are successful ongoing conference series such as *Artificial Intelligence and Theorem Proving* (AITP) and *Intelligent Computer Mathematics* (CICM) as well as workshops such as MATH-AI "The Role of Mathematical Reasoning in General Artificial Intelligence" organised within ICLR 2021. The ERC funded projects AI4REASON (2015–2020) led by Josef Urban at the Czech Technical University in Prague (CTU) and SMART (2017–2022) led by Cezary Kaliszyk at the University of Innsbruck have in particular produced a vast number of results. It is also worth noting that one of the six working groups within the new European COST Action CA20111 EuroProofNet (mentioned previously) is devoted to *Machine learning in proofs* (WG5) and led by Kaliszyk.

Though unrelated to proof assistants, it is worthwhile to mention an impressive breakthrough in machine learning for pure mathematics that has been very recently achieved thanks to a collaboration between DeepMind researchers and research mathematicians [21, 22]. More specifically, important results in knot theory [23] and representation theory [10] have been achieved by employing machine learning tools. This reinforces the hope that artificial intelligence in itself is powerful enough to provide useful auxiliary technology for research mathematics—and thus may be a very promising candidate for progress acceleration when combined with proof assistants too.

## Bibliography

[1] Aigner, M. & Ziegler, M., *Proofs from The Book*, (5th ed.), Springer-Verlag, Berlin, 2014.

[2] Anonymous, The QED Manifesto. In: Bundy, A. (ed.), *Automated Deduction. CADE-12. 12th International Conference on Automated Deduction, Nancy, France, June 26–July 1, 1994, Proceedings*. Lecture Notes in Computer Science, Vol. 814, Springer, 1994, pp. 238–251.

[3] Associate Editors of Mathematical Reviews and zbMATH, MSC2020. Mathematics Subject Classification System, 2020.

[4] Ayers, E. W., Gowers, W. T., & Jamnik, M., A Human-Oriented Term Rewriting System. In: Benzmüller C., Stuckenschmidt H. (eds.), *KI 2019: Advances in Artificial Intelligence. 42nd German Conference on AI, Kassel, Germany, September 23–26, 2019, Proceedings*. Lecture Notes in Computer Science, Vol. 11793. Springer, Cham, 2019, pp. 76–86.

[5] Baanen, A., Dahmen, S. R., Narayanan, A., & Nuccio Mortarino Majno di Capriglio, F. A. E., A Formalization of Dedekind Domains and Class Groups of Global Fields. In: Cohen, L. and Kaliszyk, C. (eds.), *12th International Conference on Interactive Theorem Proving, ITP 2021, June 29 to July 1, 2021, Rome, Italy (Virtual Conference)*. Leibniz International Proceedings in Informatics (LIPIcs), Vol. 193. Leibniz-Zentrum für Informatik, 2021, pp. 5:1–5:19.

[6] Bansal, K., Loos, S., Rabe, M., Szegedy, C., & Wilcox, S., HOList: An Environment for Machine Learning of Higher Order Logic Theorem Proving. In: Chaudhuri, K. & Salakhutdinov, R. (eds.), *Proceedings of the 36th International Conference on Machine Learning, ICML 2019, 9–15 June 2019, Long Beach, California, USA*. Proceedings of Machine Learning Research, Vol. 97, MLResearchPress, 2019, pp. 454–463.

[7] Blanchette, J. C., Kaliszyk, C., Paulson, L. C., & Urban, C., Hammering towards QED. Journal of Formalized Reasoning, 9:1, 101–148, 2016.

[8] Blanchette, J. C., Greenaway, D., Kaliszyk, C., Kühlwein, D., & Urban, J., A Learning-Based Fact Selector for Isabelle/HOL. Journal of Automated Reasoning 57, 219–244, 2016.

[9] Blanqui, F., Dowek, G., Grienenberger, É., Hondet, G., & Thiré, F., Some Axioms for Mathematics. In: Kobayashi, N. (ed.), *6th International Conference on Formal Structures for Computation and Deduction, FSCD 2021, July 17-24, 2021, Buenos Aires, Argentina (Virtual Conference)*. Leibniz International Proceedings in Informatics (LIPIcs), Vol. 195. Leibniz-Zentrum für Informatik, 2021, pp. 20:1–20:19.

[10] Blundell, C., Buesing, L., Davies, A., Veličković, P., & Williamson, G., Towards combinatorial invariance for Kazhdan-Lusztig polynomials. Preprint, 2021 (arXiv:2111.15161).

[11] Böhme, S. & Nipkow, T., Sledgehammer: Judgement Day. In: Giesl, J. & Hähnle, R. (eds.), *Automated Reasoning, 5th International Joint Conference, IJCAR 2010, Edinburgh, UK, July 16-19, 2010. Proceedings*. Lecture Notes in Computer Science Vol. 6173. Springer, 2010, pp. 107–121.

[12] Bordg, A., Paulson, L. C., & Li, W., Simple Type Theory is not too Simple: Grothendieck's Schemes without Dependent Types. Preprint, 2021 (arXiv:2104.09366).

[13] Bordg, A., Paulson, L.C., & Li, W., Grothendieck's Schemes in Algebraic Geometry. Archive of Formal Proofs, 2021, 29 March 2021.

[14] Buzzard, K., Computers and Mathematics. Newsletter of the London Mathematical Society, 484, 32–36, 2019.

[15] Buzzard, K., Commelin, J., & Massot, P., Formalising perfectoid spaces. In: Blanchette, J. & Hritcu, C. (eds.), *Proceedings of the 9th ACM SIGPLAN International Conference on Certified Programs and Proofs, CPP 2020, New Orleans, LA, USA, January 20-21, 2020*. Association for Computing Machinery, 2020, pp. 299–312.

[16] Buzzard, K., Hughes, C., Lau, K., Livingston, A., Fernández Mir, R., & Morrison, S., Schemes in Lean. To appear in: Experimental Mathematics (doi: 10.1080/10586458.2021.1983489).

[17] Buzzard, K., What is the point of computers? A question for pure mathematicians. Preprint, 2022 (arXiv:2112.11598v2).

[18] Castelvecchi, D., Mathematicians welcome computer-assisted proof in 'grand unification' theory. Nature 595, 18–19, 2021.

[19] Commelin, J., Liquid Tensor Experiment: an update. Blog post on the Lean community blog, 31 December 2021.

[20] Dahmen, S. R., Hölzl, J., & Lewis, R. Y., Formalizing the solution to the cap set problem. In: Harrison, J., O'Leary, J., & Tolmach, A. (eds.), *10th International Conference on Interactive Theorem Proving, ITP 2019, September 9–12, 2019, Portland, OR, USA*. Leibniz International Proceedings in Informatics (LIPIcs), Vol. 141. Leibniz-Zentrum für Informatik, 2019, pp. 15:1–15:19.

[21] Davies, A., Veličković, P., Buesing, L., Blackwell, S., Zheng, D., Tomašev, N., Tanburn, R., Battaglia, P., Blundell, C., Juhász, A., Lackenby, M., Williamson, G., Hassabis, D., & Kohli, P., Advancing mathematics by guiding human intuition with AI. Nature 600, 70–74, 2021.

[22] Davies, A., Kohli, P., & Hassabis, D., Exploring the beauty of pure mathematics in novel ways. Blog post on the DeepMind blog, 1 December 2021.

[23] Davies, A., Juhász, A., Lackenby, M., & Tomasev, N., The signature and cusp geometry of hyperbolic knots. Preprint, 2022 (arXiv:2111.15323v2).

[24] De Lon, A., Koepke, P., Lorenzen, A., Marti, A., Schütz, M., & Wenzel, M.. The Isabelle/Naproche Natural Language Proof Assistant. In: Platzer, A. & Sutcliffe, G. (eds.), *Automated Deduction. CADE 28. 28th International Conference on Automated Deduction, Virtual Event, July 12–15, 2021, Proceedings*. Lecture Notes in Computer Science, Vol. 12699, Springer, 2021, pp. 614–624.

[25] De Lon, A., Koepke, P., Lorenzen, A., Marti, A., Schütz, M., & Sturzenhecker, E., Beautiful Formalizations in Isabelle/Naproche. In: Kamareddine, F. & Sacerdoti Coen, C. (eds.), *Intelligent Computer Mathematics. 14th International Conference, CICM 2021, Timisoara, Romania, July 26–31, 2021, Proceedings*. Lecture Notes in Computer Science, Vol. 12833, Springer, 2021, pp. 19–31.

[26] Dillies, Y. & Mehta, B., Formalising Szemerédi's Regularity Lemma in Lean. In: Andronick, J. & de Moura, L. (eds.), *13th International Conference on Interactive Theorem Proving (ITP 2022)*. Leibniz International Proceedings in Informatics, Vol. 237, Schloss Dagstuhl, 2022, pp. 9:1–9:19.

[27] Distefano, D., Fähndrich, M., Logozzo, F., & O' Hearn, P. W., Scaling static analyses at Facebook. Communications of the ACM 62, 62–70, 2019.

[28] Doxiadis, A., *Uncle Petros and Goldbach's Conjecture*. Bloomsbury, 2000.

[29] Dunne, E. & Hulek, K., Mathematics Subject Classification 2020. EMS Newsletter, March 2020.

[30] Džamonja, M., Koutsoukou-Argyraki, A., & Paulson, L. C., Formalising Ordinal Partition Relations Using Isabelle/HOL. To appear in: Experimental Mathematics (doi: 10.1080/10586458.2021.1980464).

[31] Eberl, M., Nine Chapters of Analytic Number Theory in Isabelle/HOL. In: Harrison, J., O'Leary, J., & Tolmach, A. (eds.), *10th International Conference on Interactive Theorem Proving, ITP 2019, September 9–12, 2019, Portland, OR, USA*. Leibniz International Proceedings in Informatics (LIPIcs), Vol. 141. Leibniz-Zentrum für Informatik, 2019, pp. 6:1–6:19.

[32] Edmonds, C., Koutsoukou-Argyraki, A., & Paulson, L. C., Szemerédi's Regularity Lemma. Archive of Formal Proofs, 2021, 5 November 2021.

[33] Edmonds, C., Koutsoukou-Argyraki, A., & Paulson, L. C., Roth's Theorem on Arithmetic Progressions. Archive of Formal Proofs, 2021, 28 December 2021.

[34] Edmonds, C., Koutsoukou-Argyraki, A., & Paulson, L. C., Formalising Szemerédi's Regularity Lemma and Roth's Theorem on Arithmetic Progressions in Isabelle/HOL. Preprint, 2022 (arXiv:2207.07499).

[35] Ellenberg, J. S. & Gijswijt, D., On large subsets of $\mathbb{F}_q^n$ with no three-term arithmetic progression. Annals of Mathematics 185:1, 339–343, 2017.

[36] Erdős, P. & Milner, E. C., A theorem in the partition calculus. Canadian Mathematical Bulletin 15:4, 501–505, 1972.

[37] Erdős, P. & Straus, E. G., On the irrationality of certain series. Pacific Journal of Mathematics, 55:1, 85–92, 1974.

[38] Farah, I., All automorphisms of the Calkin algebra are inner. Annals of Mathematics, 173:2, 619–661, 2011.

[39] Ganesalingam, M. & Gowers, W. T., A fully automatic theorem prover with human-style output. Journal of Automated Reasoning 58:2, 253–291, 2017.

[40] Gouëzel, S. & Shchur, V., A corrected quantitative version of the Morse lemma. Journal of Functional Analysis, 277:4, 1258–1268, 2019.

[41] Gowers, W.T., Rough Structure and Classification. In: Alon N., Bourgain J., Connes A., Gromov M., & Milman V. (eds.), *Visions in Mathematics. GAFA 2000 Special Volume, Part I*. Modern Birkhäuser Classics. Birkhäuser, Basel, 2010, pp. 79–117.

[42] Gunther, E., Pagano, M., Sánchez Terraf, P., & Steinberg, M., The Independence of the Continuum Hypothesis in Isabelle/ZF. Archive of Formal Proofs, 2022, 6 March 2022.

[43] Hales, T. C., A proof of the Kepler conjecture. Annals of Mathematics, 162:3, 1065–1185, 2005.

[44] Hales, T., Adams, M., Bauer, G., Dang, T. D., Harrison, J., Hoang, L. T., Kaliszyk, C., Magron, V., McLaughlin, S., Nguyen, T. T., Nguyen, Q. T., Nipkow, T., Obua, S., Pleso, J., Rute, J., Solovyev, A., Ta, T. H. A., Tran, N. T., Trieu, T. D., Urban, J., Vu, K., & Zumkeller, R., A Formal Proof of the Kepler Conjecture. Forum of Mathematics, Pi, 5, e2, 2017.

[45] Han, J. M. & van Doorn, F., A Formal Proof of the Independence of the Continuum Hypothesis. Preprint, 2021 (arXiv:2102.02901).

[46] Hančl, J., Irrational rapidly convergent series. Rendiconti del Seminario Matematico della Università di Padova, 107, 225–231, 2002.

[47] Hančl, J. & Rucki, P., The transcendence of certain infinite series. Rocky Mountain Journal of Mathematics, 35:2, 531–537, 2005.

[48] Hartnett, K., Proof Assistant Makes Jump to Big-League Math. Quanta Magazine, 28 July 2021.

[49] Holden, S. B., Machine Learning for Automated Theorem Proving: Learning to Solve SAT and QSAT. Foundations and Trends in Machine Learning, 14:6, 807–989, 2021.

[50] Kjos-Hanssen, B., Niraula, S., & Yoon, S., A parametrized family of Tversky metrics connecting the Jaccard distance to an analogue of the normalized information distance. In: Artemov, S.N. & Nerode, A. (eds.), *Logical Foundations of Computer Science. International Symposium, LFCS 2022, Deerfield Beach, FL, USA, January 10–13, 2022, Proceedings*. Lecture Notes in Computer Science, Vol. 13137, Springer, 2022, pp. 112–124.

[51] Kohlenbach, U., *Applied Proof Theory. Proof Interpretations and their use in Mathematics*. Springer Monographs in Mathematics, Springer, 2008.

[52] Kohlenbach, U., Recent progress in proof mining in nonlinear analysis, IFCoLog Journal of Logics and its Applications, 10:4, 3357–3406, 2017.

[53] Kohlenbach, U., Proof-theoretic Methods in Nonlinear Analysis, In: Sirakov, B., Ney de Souza, P., & Viana, M. (eds.), *Proceedings of the International Congress of Mathematicians, Rio de Janeiro 2018. Volume II, Invited Lectures*, World Scientific, 2018, pp. 61–82.

[54] Komendantskaya, E., Heras, J., & Grov, G., Machine Learning in Proof General: Interfacing Interfaces. In: Kaliszyk, C. & Lüth, C. (eds.), *Proceedings 10th International Workshop On User Interfaces for Theorem Provers Bremen, Germany, July 11th 2012*, Electronic Proceedings in Theoretical Computer Science, Vol. 118, Open Publishing Association, 2013, pp. 15–41.

[55] Koutsoukou-Argyraki, A., Formalising Mathematics—in Praxis; A Mathematician's First Experiences with Isabelle/HOL and the Why and How of Getting Started. Jahresbericht der Deutschen Mathematiker-Vereinigung 123, 3–26, 2021.

[56] Koutsoukou-Argyraki, A. & Li, W., Irrational rapidly convergent series, Archive of Formal Proofs, 2018, 23 May 2018.

[57] Koutsoukou-Argyraki, A. & Li, W., The transcendence of certain infinite series, Archive of Formal Proofs, 2019, 27 May 2019.

[58] Koutsoukou-Argyraki, A. & Li, W., Irrationality Criteria for Series by Erdős and Straus, Archive of Formal Proofs, 2020, 12 May 2020.

[59] Koutsoukou-Argyraki, A., Li, W., & Paulson, L. C., Irrationality and Transcendence Criteria for Infinite Series in Isabelle/HOL. To appear in: Experimental Mathematics (doi: 10.1080/10586458.2021.1980465).

[60] Larson, J. A., A short proof of a partition theorem for the ordinal $\omega^\omega$. Annals of Mathematical Logic, 6, 129–145, 1973.

[61] Li, W., Yu, L., Wu, Y., & Paulson, L. C., IsarStep: a Benchmark for High-level Mathematical Reasoning. In: *9th International Conference on Learning Representations, ICLR 2021, Virtual Event, Austria, May 3-7, 2021*. OpenReview.net, 2021.

[62] de Moura, L., Kong, S., Avigad, J., van Doorn, F., & von Raumer, J., The Lean theorem prover (system description). In: Felty, A.P. & Middeldorp, A. (eds.), *Automated Deduction. CADE-25. 25th International Conference on Automated Deduction, Berlin, Germany, August 1–7, 2015, Proceedings*. Lecture Notes in Computer Science. Vol. 9195, Springer, 2015, pp. 378–388.

[63] Nash-Williams, C. St. J. A., On well-quasi-ordering transfinite sequences, Proceedings of the Cambridge Philosophical Society 61, 33–39, 1965.

[64] Nipkow, T., Paulson, L. C., & Wenzel, M., *Isabelle/HOL: A Proof Assistant for Higher-Order Logic*. Lecture Notes in Computer Science, Vol. 2283, Springer, 2002.

[65] Paulson, L. C., The Foundation of a Generic Theorem Prover. Journal of Automated Reasoning, 5:3, 363–397, 1989.

[66] Paulson, L. C., Formalising mathematics in simple type theory. In: Centrone, S., Kant, D., & Sarikaya, D. (eds.), *Reflections on the Foundations of Mathematics. Univalent Foundations, Set Theory and General Thoughts*. Synthese Library, Vol. 407, Springer, 2019, pp. 437–454.

[67] Paulson, L. C., Zermelo Fraenkel set theory in higher-order logic, Archive of Formal Proofs, 2019, 24 October 2019.

[68] Paulson, L. C., The Nash-Williams partition theorem, Archive of Formal Proofs, 2020, 16 May 2020.

[69] Paulson, L. C., Ordinal partitions, Archive of Formal Proofs, 2020, 3 August 2020.

[70] Paulson, L. C., Wetzel's Problem and the Continuum Hypothesis, Archive of Formal Proofs, 2022, 18 February 2022.

[71] Phillips, N. C., & Weaver, N., The Calkin algebra has outer automorphisms. Duke Mathematical Journal, 139:1, 185–202, 2007.

[72] Polu, S. & Sutskever, I., Generative Language Modeling for Automated Theorem Proving. Preprint, 2020 (arXiv:2009.03393v1).

[73] Scholze, P., Perfectoid spaces. Publications mathématiques de l'Institut des Hautes Études Scientifiques 116, 245–313, 2012.

[74] Scholze, P., Liquid tensor experiment. Blog post on the Xena project blog, 5 December 2020.

[75] Scholze, P., Half a year of the Liquid Tensor Experiment: Amazing developments. Blog post on the Xena project blog, 5 June 2021.

[76] Shelah, S., Infinite abelian groups, Whitehead problem and some constructions. Israel Journal of Mathematics, 18, 243–256, 1974.

[77] Stathopoulos, Y., Koutsoukou-Argyraki, A., & Paulson, L. C., SErAPIS: A Concept-Oriented Search Engine for the Isabelle Libraries Based on Natural Language. Paper published on the website of the Isabelle Workshop 2020, 2020.

[78] Stathopoulos, Y., Koutsoukou-Argyraki, A. & Paulson, L. C., Developing a Concept-Oriented Search Engine for Isabelle Based on Natural Language: Technical Challenges. Paper published on the website of the 5th Conference on Artificial Intelligence and Theorem Proving, AITP 2020, 2020.

[79] Verchinine, K., Lyaletski, A., & Paskevich, A., System for Automated Deduction (SAD): A Tool for Proof Verification. In: Pfenning, F. (ed.) *Automated Deduction. CADE-21. 21st International Conference on Automated Deduction, Bremen, Germany, July 17–20, 2007, Proceedings.* Lecture Notes in Computer Science, Vol. 4603, Springer, 2007, pp. 398–403.

[80] Whitehead, A. N. & Russell, B., *Principia Mathematica, Vol. 1, 2, 3.* Cambridge University Press, 1927.

[81] Wiedijk, F. (ed.), *The Seventeen Provers of the World.* Lecture Notes in Artificial Intelligence, Vol. 3600, Springer, 2006.

# Die Mitgliederentwicklung in der Frühzeit der DVMLG

Benedikt Löwe

Institute for Logic, Language and Computation, Universiteit van Amsterdam, Postbus 94242, 1090 GE Amsterdam, Niederlande

Fachbereich Mathematik, Universität Hamburg, Bundesstraße 55, 20146 Hamburg, Deutschland

Churchill College, Lucy Cavendish College, & Department of Pure Mathematics and Mathematical Statistics, University of Cambridge, Storey's Way, Cambridge CB3 0DS, England

E-Mail: loewe@math.uni-hamburg.de

## 1 Die Frühzeit des Vereins

Die Frühzeit der *Deutschen Vereinigung für mathematische Logik und für Grundlagenforschung der exakten Wissenschaften* (DVMLG) hat einen bleibenden Eindruck in der Vereinigung hinterlassen: die Prozeduren der frühen Vereinigung sind zu einem institutionellen Topos geworden, welcher zu verschiedenen Gelegenheiten sowohl von Zeitzeugen als auch aus zweiter Hand wiedergegeben wird.[1]

Als Frühzeit der DVMLG werden in diesem Artikel die ersten zehn Jahre (von der Gründung im Jahre 1962 bis zur grundlegenden Satzungsänderung im Jahre 1972) bezeichnet; in dieser Zeit trafen die Mitglieder jährlich bei einer Tagung in Oberwolfach zu einer Mitgliederversammlung zusammen und der Prozess der Zuwahl neuer Mitglieder war eng mit der Teilnahme an diesen Tagungen verbunden. Mit wachsender Mitgliederzahl wurde zunehmend deutlich, dass die Organisation des Vereins um die jährliche Oberwolfachtagung nicht mehr praktikabel war.[2] Nach 1973 fanden die Mitgliederversammlungen nicht mehr in Oberwolfach statt[3] und man änderte den Prozess

---

[1] Vgl. auch W. Bibel, Erinnerungen an frühe Jahre der DVMLG und D. Kant & D. Sarikaya, Gespräch mit Arnold Oberschelp, dem Vorsitzenden der DVMLG von 1970 bis 1976, in diesem Bande.

[2] Vgl. Protokoll der Mitgliederversammlung der DVMLG am 19. April 1972 in Oberwolfach (DVMLG-Archiv A56): „Herr A. Oberschelp meint, daß die Mitgliederversammlung zukünftig von der ‚Hermes-Schütte'-Tagung getrennt werden muß."

[3] Die nächsten Mitgliederversammlungen fanden am 1. August 1974 in Kiel, im Rahmen der Jahrestagung der Deutschen Mathematiker-Vereinigung am 13. Oktober 1976 in München und am 3. Oktober 1978 in Aachen statt (vgl. DVMLG-Archiv A64, A66, & A84 und Universitätsarchiv Freiburg B 0160/2.). Im Jahre 1979 gab es einen Vorstoß, die Mitgliederversammlungen wieder nach Oberwolfach zu verlegen (vgl. DVMLG-Archiv A88), dies wurde in den 1980er Jahren aber nicht umgesetzt. Vgl. auch Tabelle 2 im Vorwort dieses Bandes (S. ix).

der Aufnahme neuer Mitglieder.⁴ Am Ende des als Frühzeit bezeichneten Zeitraums waren 73 Mitglieder aufgenommen worden (von denen bis 1972 drei verstorben waren).

Der genannte institutionelle Topos beinhaltet insbesondere eine streng selektive Mitgliederpolitik: zukünftige Mitglieder mussten auf der Tagung in Oberwolfach durch einen Vortrag um die Zuwahl werben.⁵ In diesem Artikel soll dieser Topos auf der Grundlage der Unterlagen im Archiv der DVMLG beleuchtet werden. Das Archiv der DVMLG befindet sich derzeit an der Universität Hamburg und ist nur teilweise erschlossen. Die Unterlagen bis 1990 sind im Jahre 2012 von Peter Koepke und Daniel Witzke konsolidiert, in fünf Ordner (A bis E) gegliedert und teildigitalisiert worden;⁶ zusätzliche Dokumente befinden sich im Universitätsarchiv Freiburg (Archivaliensignatur B 0160/2; Laufzeit 1967–1979).

## 2 Die Vorgängerinstitution

Die Gründung der DVMLG vollzog sich in zwei Schritten: eine informelle gleichnamige Vorgängerinstitution wurde voraussichtlich im Jahre 1954 gegründet; die Gründung der DVMLG als eingetragener Verein erfolgte am 28. Juli 1962.⁷ In Abschnitt 3 wird ausgeführt, dass der eingetragene Verein DVMLG als von der Vorgängerinstitution separate juristische Person angesehen wurde.

Um die beiden Institutionen im Rahmen dieses Artikels deutlich voneinander zu unterscheiden, wird das Akronym „DVLG" für die Vorgängerinstitution und das Akronym „DVMLG" für den ab 1962 eingetragenen Verein verwendet. Diese Akronymverwendung entspricht nicht der historischen Verwendung, sondern ist lediglich eine Disambiguierungskonvention für den Kontext dieses Artikels.⁸

---

⁴Dies war bereits 1971 von Arnold Oberschelp angeregt worden: „Ich ... möchte Sie bitten zu erwägen, ob das bisherige Aufnahmeverfahren bei der wachsenden Mitgliederzahl noch optimal ist." (Brief von Arnold Oberschelp und die Vorstandsmitglieder der DVMLG v. 11. Februar 1971; DVMLG-Archiv A52). Vgl. auch B. Löwe & D. Sarikaya, Satzungen der DVMLG durch die Jahrzehnte, in diesem Bande.

⁵Heutzutage haben die meisten Fachorganisationen eine inklusive Mitgliederpolitik, bei der Personen, die sich zur Fachgemeinschaft zugehörig fühlen, durch Erklärung ihres Beitritts Mitglied werden. Ein System der Zuwahl findet sich noch in den wissenschaftlichen Akademien und in Fachgesellschaften in Großbritannien: z.B. werden die Neumitglieder der *London Mathematical Society* und der *Cambridge Philosophical Society* auf den Mitgliederversammlungen gewählt; diese Wahlen sind allerdings rein performativ und eine Ablehnung der Mitgliedschaft durch die Mitgliederversammlung ist in der Praxis undenkbar.

⁶Die Unterlagen des erschlossenen Archivs bis 1990 werden mit der Archivnummer (Ordnerbuchstabe und laufende Nummer) zitiert.

⁷DVMLG-Archiv A18.

⁸Vgl. B. Löwe & D. Sarikaya, Satzungen der DVMLG durch die Jahrzehnte, in diesem Bande: Die Satzungen beider Institutionen verwendeten das Akronym „DVLG" von 1954 bis 1967 und „DVMLG" seit 1967; im Schriftverkehr verwendete der dama-

ABBILDUNG 1. *Von links nach rechts.* Hans Hermes in Oberwolfach, ca. 1970. Paul Lorenzen in Erlangen, 1967. Werner Markwald, 1955. (Bilder: Konrad Jacobs / Konrad Jacobs / Reinhold Remmert. Quelle: Bildarchiv des Mathematischen Forschungsinstituts Oberwolfach.)

Die DVLG war ein nicht eingetragener Verein mit Satzung, Vorstand und Präsidium. Ihre Gründung erfolgte im Kontext der Vertretung der Bundesrepublik Deutschland in internationalen Wissenschaftsorganisationen, insbesondere der *Division for Logic, Methodology and Philosophy of Science of the International Union of History and Philosophy of Science* (DLMPS/IUHPS).[9] Das Archiv der DVMLG enthält nur wenige Dokumente aus der Zeit vor 1962,[10] u.a. einen Satzungsentwurf[11] und das Sitzungsprotokoll einer Sitzung am 18. November 1954 mit zehn anwesenden Mitgliedern.[12] In dieser Sitzung wurde der Satzungsentwurf besprochen und beschlossen, daß die Satzung als angenommen gilt, wenn „innerhalb eines Monats kein Einspruch erhoben wird", sowie das Präsidium „einstimmig bestätigt": es muß mindestens eine Sitzung, vermutlich die Gründungssitzung, vor dem 18. November 1954 gegeben haben, welche nicht im DVMLG-Archiv dokumentiert ist.

lige Vorsitzende Arnold Schmidt bereits 1961 das Akronym „DVMLG" in Bezug auf die Vorgängerinstitution (Brief von Schmidt an Hermes v. 5. Juni 1961; DVMLG-Archiv A14), aber auch „DVLG" in Bezug auf den eingetragenen Verein (Brief von Schmidt an Regierungsdirektor Dr. Peters v. 6. April 1964; DVMLG-Archiv A26).

[9]Für eine ausführlichere Schilderung dieser Ereignisse, vgl. B. Löwe, Grundlagenforschung der exakten Wissenschaften: die DVMLG und die Philosophie, in diesem Bande.

[10]Es ist zu erwarten, dass fehlende Dokumente aus den Jahren 1950 bis 1962 im Scholz-Nachlaß in der Universitäts- und Landesbibliothek Münster zu finden sind. Dieser ist allerdings bisher nicht vollständig erschlossen.

[11]DVMLG-Archiv A2; vgl. auch B. Löwe & D. Sarikaya, Satzungen der DVMLG durch die Jahrzehnte, in diesem Bande.

[12]DVMLG-Archiv A3.

ABBILDUNG 2. *Von links nach rechts.* Arnold Schmidt in Oberwolfach, 1949. Kurt Schütte in München, ca. 1985. Walter Oberschelp in Oberwolfach. (Bilder: Universitätsarchiv Freiburg, Depositalbestand E6 / Gerd Fischer / Konrad Jacobs. Quelle: Bildarchiv des Mathematischen Forschungsinstituts Oberwolfach.)

Aus den Jahren 1954 bis 1959 gibt es keine Dokumente im DVMLG-Archiv. Bezüglich der an die DLMPS/IUHPS zu zahlenden Mitgliedsbeiträge gab es in den Jahren 1959 bis 1961 einen Briefverkehr zwischen dem Vorsitzenden Arnold Schmidt, der Deutschen Forschungsgemeinschaft (DFG), der DLMPS/IUHPS und dem Bundesinnenministerium, in dem die Zahlungsmodalitäten dieser Beiträge verhandelt wurden. Die Bitte des Bundesinnenministeriums nach einem Haushaltsplan legte eine formalere Struktur und die Eintragung der DVLG ins Vereinsregister nahe:[13] „[E]s wäre wohl eine Mitgliederversammlung der bisherigen Mitglieder erforderlich, die das alles beschließt".[14] Es ist zu vermuten, daß Schmidt und Hermes sich im Herbst 1961 getroffen haben und daß Schmidt im Winter 1961/62 mit der DFG technische Einzelheiten besprochen hat.[15] Am 28. Juli 1962 wurde die DVMLG in Marburg an der Lahn gegründet (s. Abschnitt 3).

Im Archiv der DVMLG liegt keine Mitgliederliste der DVLG vor, so dass die Mitglieder dieser Vorgängerinstitution nur indirekt ermittelt wer-

---

[13]Für eine detailliertere Diskussion, s. B. Löwe & D. Sarikaya, Satzungen der DVMLG durch die Jahrzehnte, in diesem Bande, § 2.

[14]Brief von Schmidt an Hermes v. 5. Juni 1961 (DVMLG-Archiv A14).

[15]Schmidt schreibt am 31. Juli 1961 an Hermes: „Falls Sie auf Ihrer eventuellen Reise in den Süden hier in Marburg vorbeikommen könnten, wäre das natürlich am besten (DVMLG-Archiv A15)". Zwei Briefe von der DFG erwähnen den möglichen Besuch von Schmidt in Bonn: Brief von Dr. Müller-Daehn (DFG) an Schmidt vom 2. August 1961 (DVMLG-Archiv A16); Brief von Dr. Müller-Daehn (DFG) an Schmidt vom 6. November 1961 (DVMLG-Archiv A17).

den können. Das Sitzungsprotokoll vom 18. November 1954 listet zehn anwesende Mitglieder auf: „Prof. Dr. Ackermann, Prof. Dr. Behmann, Prof. Dr. Hermes, Prof. Dr. Lorenzen, Prof. Dr. Arnold Schmidt, Dr. Hasenjäger [*sic!*], Dr. v. Kempski, Dr. Markwald, Dr. Schütte und Dr. Wette" und berichtet, daß „Herr Dr. Markwald ... einstimmig als Mitglied aufgenommen" und daß „das Präsidium (Prof. Dr. Hermes, Prof. Dr. Arnold Schmidt und Prof. Dr. Scholz) ... einstimmig bestätigt" wurde. Es ist daher davon auszugehen, dass Heinrich Scholz bei der im DVMLG-Archiv nicht dokumentierten Gründungssitzung der DVLG anwesend gewesen ist. Im Jahre 1959 schickte Schmidt ein „vorläufiges Memorandum" an die DFG, in dem Hintergrundinformation über die DVLG gegeben wird:

> [Der DVLG] gehören unter anderem die führenden Forscher und Universitätslehrer dieses Gebietes—wie Prof. Dr. W. Ackermann, Münster; Prof. Dr. W. Britzlmayer [*sic!*], München; Prof. Dr. G. Hasenjäger [*sic!*], Münster; Prof. Dr. H. Hermes, Münster; Prof. Dr. P. Lorenzen, Kiel; Prof. Dr. H. Arnold Schmidt, Marburg a. d. Lahn; Prof. Dr. K. Schütte, Marburg a. d. Lahn—an.[16]

Aus diesen beiden Dokumenten sind also zwölf Mitglieder der DVLG gesichert: in alphabetischer Reihenfolge Ackermann, Behmann, Britzelmayr, Hasenjaeger, Hermes, von Kempski, Lorenzen, Markwald, Schmidt, Scholz, Schütte und Wette.[17]

## 3 Die DVMLG als separate Institution

Wie berichtet, hatte Schmidt im Jahre 1961 den Plan gefasst, auf einer Mitgliederversammlung der DVLG die Eintragung des Vereins ins Vereinsregister zu beschliessen. Dieser Plan wurde so nicht in die Tat umgesetzt; stattdessen wurde eine neue, gleichnamige Institution gegründet: mindestens vier Mitglieder der DVLG sind zu keinem Zeitpunkt Mitglieder der neuen Institution geworden: Heinrich Behmann, Wilhelm Britzlmayr, Heinrich Scholz und Eduard Wette.[18] Das deutlichste Zeichen für die klare Trennung der beiden Institutionen ist die Mitgliedschaft von Werner Markwald: wie berichtet, ist die Mitgliedschaft Markwalds in der DVLG durch das Sitzungsprotokoll vom 18. November 1954 bestätigt; dennoch wurde er auf der

---

[16] Brief von Schmidt an Dr. Treue (DFG) vom 17. Dezember 1959 & Vorläufiges Memorandum (DVMLG-Archiv A4).

[17] Das vermutlich erste Dokument zur Gründungsgeschichte der DVLG ist ein Brief von Heinrich Scholz an Ackermann, Behmann, Hermes, von Kempski, Lorenzen, Schmidt, Karl Schröter und Schütte vom 3. August 1950, in dem die Adressaten gefragt werden, ob sie bereit sind, an der Gründung der DVLG mitzuwirken (Scholz-Nachlaß, Universitäts- und Landesbibliothek Münster). Allerdings fand die tatsächliche Gründung der DVLG erst fast vier Jahre später statt, so dass der Mitgliedsstatus von Karl Schröter hieraus nicht eindeutig ermittelt werden kann.

[18] Scholz war am 30. November 1956 verstorben.

ABBILDUNG 3. *Links.* Dieter Rödding in Münster bei der Emeritierung Heinrich Behnkes am 4. März 1967. *Rechts.* Ernst Thiele in Oberwolfach, 1987. (Bilder: Konrad Jacobs. Quelle: Bildarchiv des Mathematischen Forschungsinstituts Oberwolfach.)

Mitgliederversammlung am 16. April 1964 formal als Mitglied der DVMLG zugewählt.[19]

Die Gründungsmitglieder der DVMLG sind Wilhelm Ackermann, Gisbert Hasenjaeger, Hans Hermes, Jürgen von Kempski, Paul Lorenzen, Arnold Schmidt und Kurt Schütte, von denen Hasenjaeger, Hermes, Schmidt und Schütte bei der Gründungssitzung in Marburg an der Lahn anwesend waren; die anderen drei hatten Schmidt eine schriftliche Vollmacht gegeben, um die durch § 59 BGB vorgegebene Mindestzahl von sieben Mitgliedern zu erfüllen.[20]

Nach ihrer Gründung hatte die DVMLG zunächst diese Mindestzahl von sieben Mitgliedern, allesamt Mitglieder des siebenköpfigen Vorstandes; am 24. Dezember 1962 verstarb Ackermann, wodurch die Zahl der Mitglieder auf sechs sank. Aus den Archivunterlagen ist nicht zu erkennen, daß es zwischen Gründung und April 1964 Vereinsaktivitäten außer den institutionellen Formalia (Eintragung ins Vereinsregister und Fortsetzung der Diskussion mit der DLMPS um die Jahresbeiträge des deutschen Nationalkomitees) gab; insbesondere gab es vor April 1964 keine Logik-Tagungen in Oberwolfach (s. Abschnitt 4).

---

[19] DVMLG-Archiv A27.
[20] DVMLG-Archiv A18.

Frühe Mitgliederentwicklung

ABBILDUNG 4. *Links.* Friedrich Bachmann in Erlangen, 1969. *Rechts.* Walter Felscher in Oberwolfach, 1976. (Bilder: Konrad Jacobs. Quelle: Bildarchiv des Mathematischen Forschungsinstituts Oberwolfach.)

Bei der nächsten Mitgliederversammlung am 16. April 1964 in Oberwolfach waren alle sechs noch lebenden Mitglieder anwesend. Das handschriftliche Protokoll berichtet:

> Auf Beschluß der Mitgliederversammlung wurden als neue Mitglieder zugewählt die Herren Müller, A. Oberschelp, W. Oberschelp, Rödding, Thiele, Lorenz, Markwald, [Mitglied Nr. 10].[21] Aufgefordert als Mitglieder beizutreten sollen die Herren Stegmüller, [Mitglied Nr. 17], Bachmann, Britzelmayer [sic!], ([aus Datenschutzgründen nicht namentlich genannt]), Gericke.[22]

Es gibt keine Archivdokumentation der Beitrittsaufforderungen an die genannten Personen oder ihrer Antworten. Das Protokoll der folgenden Mitgliederversammlung vom 8. April 1965 enthält eine Liste der Anwesenden sowie eine Liste der Entschuldigungen, woraus man eine vollständige Mitgliederliste mit Stand vom April 1965 ermitteln kann: zusätzlich zu den acht am 16. April 1969 explizit zugewählten Mitgliedern sind Bachmann,

---

[21]Die datenschutzrechtliche Behandlung von persönlichen Daten in diesem Artikel wird am Anfang von Abschnitt 7 erläutert. Eigennamen, welche aus datenschutzrechtlichen Gründen nicht genannt werden, sind durch die Mitgliedsnummern in der Liste aus Abschnitt 7 ersetzt worden.

[22]DVMLG-Archiv A27. Unter von Kempskis Unterschrift als Protokollführer findet sich die Bemerkung: „Nach den Aufzeichnungen von Herrn Prof. A. Schmidt wurde die Aufforderung an [die in Klammern genannte Person] zurückgestellt."

Stegmüller und Mitglied Nr. 17 der Einladung gefolgt und werden im Jahre 1965 als Mitglieder aufgelistet.[23]

## 4 Der Vortrag in Oberwolfach

Von 1964 bis 1972 fanden die jährlichen Mitgliederversammlungen der DVMLG im Rahmen der jährlichen Logiktagungen am Mathematischen Forschungsinstitut Oberwolfach (MFO) statt.[24] Die relevanten Logiktagungen im genannten Zeitraum sind im folgenden aufgelistet; die Teilnehmer- und Vortragslisten dieser Tagungen finden sich in *Gästebuch 2*, *Gästebuch 3*, sowie den *Vortragsbüchern* 8, 9, 11, 12, 14, 16, 19 und 21,[25] welche die Rekonstruktion der Vorträge der Mitgliedschaftskandidaten und Mitglieder ermöglichen.

(1) *Kolloquium der Deutschen Vereinigung für Mathematische Logik und für Grundlagenforschung der exakten Wissenschaften, 14. IV–17. IV 1964.*[26]

(2) *Tagung über Cohen's Unabhängigkeitsbeweise i. d. Mengenlehre, 2.IV.65–5.IV.65.* und *6. Kolloquium der DVMLG, 6.IV.65–9.IV.65.*[27]

(3) *Zur mathematischen Logik, 3.–8. April 1967.*[28]

(4) *Tagung zur mathematischen Logik 1.4.68–6.4.68.*[29]

(5) *Tagung zur Mathematischen Logik, 23.3.–29.3.1969.*[30]

(6) *Tagung zur Mathematischen Logik (5.4.–12.4.1970).*[31]

(7) *Tagung über Mathematische Logik 29.3.–3.4.1971.*[32]

(8) *Tagung über Mathematische Logik 16.4.–22.4.1972.*[33]

---

[23] DVMLG-Archiv A38.

[24] Eine Ausnahme war die Mitgliederversammlung am 9. August 1966 im Rahmen des internationalen *Kolloquium über Logik und Grundlagen der Mathematik* in Hannover (DVMLG-Archiv A40 & A41).

[25] Oberwolfach Digital Archive (ODA) E20/00149, E20/00009, E20/00010, E20/00012, E20/00013, E20/00015, E20/00018, E20/00020 und E20/00022.

[26] ODA E20/00149, Blatt 164. ODE E20/00009, Blatt 46–49.

[27] ODA E20/00149, Blatt 178–179. ODA E20/00010, Blatt 48–66. Es ist aufgrund der Archivunterlagen nicht rekonstruierbar, warum dieses Kolloquium als das „sechste" bezeichnet wurde.

[28] ODA E20/00149, Blatt 227. ODA E20/00012, Blatt 97–107.

[29] ODA E20/00149, Blatt 256. ODA E20/00013, Blatt 169–185.

[30] ODA E20/00149, Blatt 290–291. ODA E20/00015, Blatt 128–143.

[31] ODA E20/00149, Blatt 340. ODA E20/00018, Blatt 1–25.

[32] ODA E20/00150, Blatt 43. ODA E20/00020, Blatt 37–56.

[33] ODA E20/00150, Blatt 92. ODA E20/00022, Blatt 242–259.

ABBILDUNG 5. *Von links nach rechts.* Karl-Heinz Diener in Berkeley, 1973. Justus Diller in Memmingen, 1971. Heinz-Dieter Ebbinghaus in Berkeley, 1974. (Bilder: George M. Bergman / Konrad Jacobs / George M. Bergman. Quelle: Bildarchiv des Mathematischen Forschungsinstituts Oberwolfach.)

Die erwähnte erste Sitzung der DVMLG nach der Gründung fand am 16. April 1964 im Rahmen der als (1) gelisteten Tagung statt. In Abschnitt 3 wurde berichtet, daß acht Personen explizit zugewählt wurden und fünf weitere zum Beitritt aufgefordert wurden, von denen drei dieser Aufforderung nachgekommen sind. Von diesen dreizehn Personen waren lediglich vier Personen anwesend und nur drei von Ihnen haben bei der Tagung einen Vortrag gehalten.[34] Eine Erwartung oder Verpflichtung von Neumitgliedern, einen Vortrag zu halten, gab es also bei dieser ersten Zuwahl noch nicht.

Der Pflichtvortrag wurde in der Satzungsänderung im Jahre 1967 in der Satzung verankert:

> Ein für die Aufnahme vorgeschlagener Kandidat soll anläßlich einer Mitgliederversammlung vor der Abstimmung über die Aufnahme einen wissenschaftlichen Vortrag halten, Ausnahmen hiervon müssen im Antrag begründet werden und bedürfen der Zustimmung von zwei Dritteln der Anwesenden Mitglieder.[35]

---

[34] Kuno Lorenz, Arnold Oberschelp und Dieter Rödding; vgl. ODA E20/00009, Blatt 46 & 47. Kuno Lorenz schreibt: „Es war im Zusammenhang einer Arbeitstagung der DVMLG in Oberwolfach, an der auch Paul Lorenzen teilnahm und diese Gelegenheit nutzte, um mich ebenfalls dort mit meiner Arbeit zu einem dialogischen—also spieltheoretischen—Zugang zur Logik vorzustellen, daß ich Mitglied der DVMLG (und der Skatrunde ebendort mit Hans Hermes, Gert Müller und Arnold Schmidt) wurde. Mittlerweile ist ein halbes Jahrhundert vergangen und die Jüngsten von damals sind—sofern sie noch leben—die Veteranen von heute geworden." (Persönliche Kommunikation; 30. März 2022.)

[35] DVMLG-Archiv A43.

ABBILDUNG 6. *Von links nach rechts.* Heinz Gumin bei der Verleihung des *Oberwolfach Prize 2003* in Oberwolfach im Jahre 2004. Ernst Specker, 1982. Ulrich Felgner, 2011. (Bilder: Renate Schmid / Konrad Jacobs / Cornelia Niederdrenk-Felgner. Quelle der ersten beiden Bilder: Bildarchiv des Mathematischen Forschungsinstituts Oberwolfach.)

Bereits bei der nächsten, vom damaligen Präsidenten Arnold Oberschelp betriebenen Satzungsänderung im Jahre 1972 wird die Vortragspflicht wieder aus der Satzung entfernt.[36]

## 5 Zuwahlen 1965 bis 1972

Nach der Tagung im Jahre 1964 hatte die Vereinigung achtzehn Mitglieder, von denen siebzehn noch lebten, davon sechs Gründungsmitglieder, acht explizit zugewählte und drei zum Beitritt aufgeforderte. Im folgenden werden die Zuwahlen bei den nächsten acht Mitgliederversammlungen, bei denen insgesamt 55 weitere Mitglieder zugewählt wurden, im einzelnen diskutiert (vgl. auch Abschnitt 7).

**1965.** Auf der Mitgliederversammlung am 8. April 1965 in Oberwolfach wurden vier Mitglieder zugewählt, die alle auf der entsprechenden Tagung vorgetragen haben.[37] Die Zahl der Mitglieder steigt auf einundzwanzig.

**1966.** Auf der Mitgliederversammlung am 9. August 1966 in Hannover wurden vier neue Mitglieder zugewählt. Von diesen hatte zu diesem Zeitpunkt noch keiner in Oberwolfach vorgetragen; zwei der vier stehen im Pro-

---

[36]Vgl. B. Löwe & D. Sarikaya, Satzungen der DVMLG durch die Jahrzehnte, in diesem Bande.
[37]DVMLG-Archiv A38. ODA E20/00010, Blatt 48, 59–60, 63 & 65.

gramm der Tagung in Hannover.[38] Nachdem die Versammlung bereits drei Mitglieder gewählt hat, wird der Wahlvorgang unterbrochen:

> Nachdem Herr Oberschelp [eine weitere Person, Mitglied Nr. 27; vgl. Abschnitt 7] vorgeschlagen hatte, gab der Vorsitzende zu bedenken, ob es nicht besser sei, die vorgeschlagenen Herren zunächst zu einer Tagung der DVMLG einzuladen, damit auch die anderen Mitglieder diese Herren vorher kennenlernen können.[39]

Man einigt sich, zwei weitere Kandidaten auf die nächste Oberwolfach-Tagung einzuladen. Ein Sonderfall ist ein ausländischer Kandidat, der „mit der besonderen Begründung [vorgeschlagen wird], daß [er] nur noch dieses Jahr seinen Wohnsitz in Deutschland habe ... und daher nach der Satzung wählbar sei." Dieser Kandidat wird ebenfalls zugewählt (Mitglied Nr. 26; vgl. Abschnitt 7). Die Gesamtzahl der Mitglieder steigt auf fünfundzwanzig.

**1967.** Auf der Mitgliederversammlung am 6. April 1967 in Oberwolfach wurden sechs Mitglieder zugewählt, die alle bei der entsprechenden Tagung vorgetragen haben. Darunter befindet sich auch eine der in Abschnitt zum Jahre 1966 genannten Personen, die eingeladen wurden, „damit auch die anderen Mitglieder [sie] vorher kennenlernen können" (Mitglied Nr. 27).[40] Die Zahl der Mitglieder steigt auf einunddreißig; allerdings verstirbt Arnold Schmidt am 16. September 1967 und die Mitgliederzahl sinkt damit auf dreißig.

**1968.** Auf der Mitgliederversammlung am 4. April 1968 in Oberwolfach wurden fünf Mitglieder zugewählt.[41] Zwei dieser fünf hatten in diesem Jahre oder früher bereits in Oberwolfach vorgetragen, aber die anderen drei hatten bislang keinen Vortrag gehalten. Das Protokoll der Mitgliederversammlung gibt keinen Hinweis für die Gründe oder ob dieser Sachverhalt diskutiert wurde. Die Mitgliederzahl steigt auf fünfunddreißig.

**1969.** Auf der Mitgliederversammlung am 27. März 1969 in Oberwolfach wurden zehn Mitglieder zugewählt.[42] Von diesen zehn haben sechs auf der entsprechenden Tagung vorgetragen und zwei hatten bereits im Jahre 1968 vorgetragen. Eine weitere Person ist als Teilnehmer, aber nicht als Vortragender bei der Tagung im Jahr 1965 aufgelistet. Bei der verbleibenden Person

> wurde beschlossen, von der Sollbestimmung aus § 3 der Satzung, daß ein aufzunehmendes Mitglied auf einer mit einer Mitgliederversammlung verbundenen Tagung einen Vortrag halten soll, abzusehen.

---

[38] Beide trugen am 12. August, also drei Tage nach der Sitzung, vor; vgl. DVMLG-Archiv A40.
[39] DVMLG-Archiv A41.
[40] DVMLG-Archiv A43. ODA E20/00012, Blatt 98, 100, 102–103 & 105.
[41] DVMLG-Archiv A48.
[42] DVMLG-Archiv A49.

Dies ist die einzige explizite Erwähnung eines Abweichens von der Sollbestimmung. Die Mitgliederzahl steigt auf fünfundvierzig.

**1970.** Auf der Mitgliederversammlung am 9. April 1970 werden neun neue Mitglieder zugewählt.[43] Von diesen haben acht in diesem oder einem früheren Jahr in Oberwolfach vorgetragen; das verbleibende Neumitglied wird in den Vortragsbüchern erst im Jahre 1971 erwähnt.[44] Die Mitgliederzahl steigt auf vierundfünfzig.[45]

**1971.** Laut Protokoll der Mitgliederversammlung vom 1. April 1971 werden elf neue Mitglieder zugewählt, darunter Paul Bernays als Ehrenmitglied.[46] Eines der neu gewählten Mitglieder, Béla Johos, verstarb kurz nach der Tagung, am 27. Mai 1971. Die Zahl der Mitglieder steigt auf vierundsechzig.[47]

**1972.** Laut Protokoll der Mitgliederversammlung vom 19. April 1972 werden sechs neue Mitglieder zugewählt.[48] Alle sechs haben in diesem oder einem früheren Jahr in Oberwolfach vorgetragen. Das Protokoll erwähnt explizit: „Damit ist die Mitgliederzahl auf 70 angestiegen." Dies wird von der archivierten Mitgliederliste bestätigt.[49]

## 6 Zusammenfassung

Trotz eines selektiven Prinzips der Mitgliederzuwahl hat es keine Phase der künstlichen Beschränkung der Mitgliederzahl gegeben: auf der ersten Sitzung nach der Gründungssitzung des Vereins verdreifachte sich die Mitgliederzahl und weitere Zuwahlen führen in jedem folgenden Jahr zu einer substantiellen Vergrößerung des Vereins.

Den Pflichtvortrag in Oberwolfach, der als wichtigstes Instrument der Selektivität Teil des institutionellen Topos ausmacht, gab es lediglich zwischen 1967 und 1972. Dieses Prinzip wird erstmals auf der Mitgliederversammlung im Jahre 1966 diskutiert, in die neue Satzung aufgenommen und dann in den sechs Mitgliederversammlungen zwischen 1967 und 1972 angewandt. Auch in diesen Jahren wird das Aufnahmekriterium nicht immer ausführlich diskutiert: zwischen 1967 und 1972 gibt es insgesamt sieben Mitglieder, die gemäß Vortragsbüchern des MFO nicht vor ihrer Zuwahl in Oberwolfach vorgetragen haben; nur bei einem dieser Zuwahlverfahren wird dies explizit

---

[43] DVMLG-Archiv A51.
[44] ODA E20/00013, Blatt 175.
[45] Dies entspricht exakt der Mitgliederliste Stand April 1970 aus dem Archiv (DVMLG-Archiv E3).
[46] DVMLG-Archiv A54. Die Satzung der DVMLG sieht keine Ehrenmitglieder vor: die Mitgliederversammlung definiert diesen Status und seine Bedeutung *ad hoc*: „Als [Ehrenmitglied] braucht er keinen Vereinsbeitrag zu entrichten und wird jedes Jahr zur Tagung der DVMLG eingeladen."
[47] DVMLG-Archiv E4.
[48] DVMLG-Archiv A56.
[49] DVMLG-Archiv E5.

im Protokoll der Mitgliederversammlung vermerkt und eine Ausnahme von der Soll-Bestimmung beschlossen.

Es gibt keine Belege im Archiv, daß es bei der Zuwahl Diskussionen um die Inhalte der Vorträge gegeben hätte und der Pflichtvortrag somit als inhaltliches Kriterium für die Entscheidung über die Zuwahl verwendet wurde.

## 7 Liste der Mitglieder der Frühzeit der DVMLG

Im folgenden nennen wir die ersten dreiundsiebzig Mitglieder der DVMLG bis zum 19. April 1972. Bei gleichem Aufnahmedatum werden die Mitglieder in der Reihenfolge aufgelistet, in der ihre Namen im entsprechenden Protokoll der Mitgliederversammlung erwähnt werden. Aus Datenschutzgründen werden nur Mitglieder, die eine explizite Zustimmung erteilt haben oder bei denen die archivalische Schutzfrist nach §11 (2) BArchG abgelaufen ist, namentlich genannt. Die anderen Mitglieder werden mit laufender Nummer aufgelistet. Alle Lebensdaten sind entweder mit expliziter Zustimmung der oder des Betroffenen veröffentlicht oder sind öffentlich zugänglichen Quellen entnommen (z.B. Jahrbücher, öffentliche Nachrufe, öffentlich zugängliche Archive von E-Mail-Listen etc.). Die Quellenangaben beziehen sich auf das DVMLG-Archiv.

**Gründungsmitglieder: 28. Juli 1962 (A18).**

1. Gisbert Hasenjaeger, 1. Juni 1919 – 2. September 2006.
2. Johann (Hans) Hermes, 12. Februar 1912 – 10. November 2003.
3. H. Arnold Schmidt, 11. Juli 1902 – 16. September 1967.
4. Kurt Schütte, 14. Oktober 1909 – 18. August 1998.
5. Wilhelm Ackermann, 29. März 1896 – 24. Dezember 1962.
6. Jürgen von Kempski, 20. Mai 1910 – 11. Oktober 1998.
7. Paul Lorenzen, 24. März 1915 – 1. Oktober 1994.

**Zuwahl 16. April 1964 (A27 & A38).**

8. Gert H. Müller, 29. Mai 1923 – 3. September 2006.
9. Arnold Oberschelp, geboren 5. Februar 1932.
10.
11. Dieter Rödding, 24. August 1937 – 4. Juni 1984.
12. Ernst Thiele, verstorben 2001.
13. Kuno Lorenz, geboren 17. September 1932.
14. Werner Markwald, 1925 – 18. Mai 1973.
15. Verstorben 2015.
16. Wolfgang Stegmüller, 3. Juni 1923 – 11. Juni 1991.
17. Verstorben 2018.
18. Friedrich Bachmann, 11. Februar 1909 – 1. Oktober 1982.

**Zuwahl 8. April 1965 (A38).**

**19.** Wolfram Schwabhäuser, 20. Mai 1931 – 27. Dezember 1985.
**20.**
**21.** Walter Felscher, 12. Oktober 1931 – 9. Dezember 2000.
**22.**

**Zuwahl 9. August 1966 (A41).**

**23.** Justus Diller, geboren 6. Juni 1936.
**24.** Helmut Pfeiffer, verstorben 2. Juli 2004.
**25.** Karl-Heinz Diener, verstorben 18. September 2007.
**26.** Verstorben 2019.

**Zuwahl 6. April 1967 (A43).**

**27.**
**28.** Verstorben 2022.
**29.** Heinz-Dieter Ebbinghaus, geboren 22. Februar 1939.
**30.**
**31.**
**32.** Martin Löb, 31. März 1921 – 21. August 2006.

**Zuwahl 4. April 1968 (A48).**

**33.** Heinz Gumin, 19. August 1928 – 24. November 2008.
**34.** Verstorben 2020.
**35.** Verstorben 2014.
**36.** Klaus Potthoff, geboren 26. September 1942.
**37.** Jürgen Schmidt, 5. August 1918 – 14. Oktober 1980.

**Zuwahl 27. März 1969 (A49).**

**38.** Ernst Specker, 11. Februar 1920 – 10. Dezember 2011.
**39.** Verstorben 2019.
**40.**
**41.** Ulrich Felgner, geboren 17. November 1941.
**42.** Sabine Koppelberg, geb. Görnemann.
**43.** Albert Menne, 12. Juli 1923 – 7. März 1990.
**44.**
**45.** Verstorben 2020.
**46.** Helmut Schwichtenberg, geboren 5. April 1942.
**47.** Christian Thiel, geboren 12. Juni 1937.

**Zuwahl 9. April 1970 (A51).**

48. Wolfgang Bibel, geboren 28. Oktober 1938.
49. Jörg Flum.
50. Verstorben 2016.
51. Verstorben 2022.
52.
53. Verstorben 2019.
54.
55. Alexander Prestel, geboren 1941.
56. Verstorben 2016.

**Zuwahl 1. April 1971 (A54).**

57.
58.
59. Gerhardt Frey, 19. Oktober 1915 – 19. Juni 2002
60.
61. Béla Juhos, 22. November 1901 – 27. Mai 1971.
62. Hans Läuchli, 1933 – 1997.
63.
64.
65.
66. Claus Peter Schnorr, geboren 4. August 1943
67. Paul Bernays, 17. Oktober 1888 – 18. September 1977 (*Ehrenmitglied*).

**Zuwahl 19. April 1972 (A56).**

68. Bruno Buchberger, geboren 22. Oktober 1942.
69. Egon Börger, geboren 13. Mai 1946.
70.
71. Roland Fraïssé, 3. Dezember 1920 – 30. März 2006.
72. Verstorben 2019.
73. Martin Ziegler.

**Zusätzliche DVLG-Mitglieder in alphabetischer Reihenfolge.**

1. Heinrich Behmann, 10. Januar 1891 – 3. Februar 1970.
2. Wilhelm Britzelmayr, 27. August 1892 – 1970.
3. Heinrich Scholz, 17. Dezember 1884 – 30. Dezember 1956.
4. Eduard Wette, geboren 4. Februar 1925; verstorben.

# Grundlagenforschung der exakten Wissenschaften: die DVMLG und die Philosophie

## Benedikt Löwe*

Institute for Logic, Language and Computation, Universiteit van Amsterdam, Postbus 94242, 1090 GE Amsterdam, Niederlande

Fachbereich Mathematik, Universität Hamburg, Bundesstraße 55, 20146 Hamburg, Deutschland

Churchill College, Lucy Cavendish College, & Department of Pure Mathematics and Mathematical Statistics, University of Cambridge, Storey's Way, Cambridge CB3 0DS, England

E-Mail: loewe@math.uni-hamburg.de

Die Satzung der DVMLG verweist auf die „in ihrem Namen aufgeführten Wissenschaftszweige"; viele Mitglieder verstehen den eponymen Begriff *Mathematische Logik und Grundlagenforschung der exakten Wissenschaften* (verkürzt oft *Logik und Grundlagenforschung*) als Hendiadyoin und fassen das so benannte Gebiet als Teilgebiet der Mathematik auf. Dieser Artikel beleuchtet den Hintergrund der Gründung und Namensgebung der DVMLG und erörtert die folgenden Fragen:

> War das Gebiet *Logik und Grundlagenforschung* im Verständnis der Gründer der DVMLG ein einheitliches Gebiet oder zwei separate Forschungszweige und wurde es als Teilgebiet der Mathematik verstanden?

Quellen für diesen Artikel sind das Archiv der DVMLG und Teile des Scholz-Nachlasses in der Universitäts- und Landesbibliothek Münster. Das Archiv der DVMLG befindet sich derzeit an der Universität Hamburg; die Unterlagen bis 1990 sind im Jahre 2012 von Peter Koepke und Daniel Witzke konsolidiert, in fünf Ordner (A bis E) gegliedert und teildigitalisiert worden. Es befinden sich zusätzliche Dokumente im Universitätsarchiv Freiburg (Archivaliensignatur B 0160/2; Laufzeit 1967–1979). Der Scholz-Nachlaß in der Universitäts- und Landesbibliothek Münster ist bislang nur teilweise erschlossen.[1]

---

*Der Autor dankt Volker Peckhaus, Andrea Reichenberger und Niko Strobach für Diskussionen über Heinrich Scholz. Besonderer Dank gilt Andrea Reichenberger, Jürgen Lenzing und den involvierten Mitarbeiterinnen und Mitarbeitern der Universitäts- und Landesbibliothek Münster für das Auffinden und Bereitstellen von Dokumenten aus dem Scholz-Nachlaß.

[1] Dokumente aus dem erschlossenen Teil des Scholz-Nachlasses werden mit Kapselnummer und Dokumentennummer zitiert; Dokumente aus dem nicht erschlossenen Teil werden lediglich als „Scholz-Nachlaß" zitiert.

ABBILDUNG 1. Heinrich Scholz und Hans Hermes, 1952. (Bild: Reinhold Remmert. Quelle: Bildarchiv des Mathematischen Forschungsinstituts Oberwolfach.)

## 1 Heinrich Scholz und die DVMLG

Die DVMLG wurde zweimal gegründet, zunächst als nichteingetragener Verein, vermutlich im Jahre 1954,[2] dann als eingetragener Verein in Marburg am 28. Juli 1962.[3] Dieser Gründungsprozess ist unmittelbar und eng mit der Vertretung des Gebiets *Logik und Grundlagenforschung* in der damals neuen globalen Wissenschaftslandschaft verknüpft: der Anlaß für die erste Gründung war die Bitte Ferdinand Gonseths (1890–1975), eine „deutsche Sektion" der *Societé International de Logique et de Philosophie des Sciences* (SILPS) zu gründen. Dieser Hintergrund wird in Abschnitt 2 ausführlich geschildert.

Eine der zentralen Personen in diesem Prozess (und Adressat der erwähnten Bitte Gonseths) war Heinrich Scholz (1884–1956) aus Münster, der ausschlaggebend für die Wahl des Namens der Vereinigung war. Scholz war ein protestantischer systematischer Theologe, bekannt für sein Werk

---

[2] Das Archiv der DVMLG enthält das Protokoll einer Sitzung des nichteingetragenen Vereins in Marburg am 18. November 1954. Diese Sitzung ist nicht die Gründungssitzung, aber das erste Präsidium (Hans Hermes, 1912–2003, Arnold Schmidt, 1902–1967, und Heinrich Scholz) wurde bestätigt und ein Satzungsentwurf wurde diskutiert; es ist davon auszugehen, dass es sich um die erste Sitzung nach der Gründung der Vereinigung handelt (DVMLG-Archiv A3). Vgl. auch (Löwe, 2022).

[3] DVMLG-Archiv A18.

zur Religionsphilosophie (Scholz, 1921), der, von Russell und Whitehead beeinflußt, in das Gebiet der mathematischen Logik wechselte. Scholz war fasziniert von der Anwendung rein mathematischer Methoden auf Fragen der philosophischen Grundlagen und war davon überzeugt, daß es sich bei diesem innovativen Zugang zu den Grundlagen der Mathematik um ein transformatives, zukunftsweisendes Forschungsgebiet handelt.

Er wurde 1928 auf einen Lehrstuhl für Philosophie nach Münster berufen und betrieb in den folgenden Jahren die Umorientierung dieses Lehrstuhls. Zwischen 1928 und 1938 versuchte Scholz einen geeigneten Namen für dieses Forschungsgebiet zu finden, um es als wichtiges Forschungsgebiet in Deutschland zu etablieren. In dieser Zeit wurde das Wort „Logistik" im deutschen Sprachraum synonym mit „formale Logik" oder „symbolische Logik" verwendet und Scholz' Forschungsgebiet wurde zunächst als „logistische Logik und Grundlagenforschung" bezeichnet und seine Abteilung als „Logistische Abteilung des Philosophischen Seminars".[4] Scholz bevorzugte stattdessen die Begriffskomplexe „neue mathematische Logik und Grundlagenforschung", „mathematische Logik und Grundlagenforschung" und „Logik und Grundlagenforschung".[5] Im Folgenden wird in diesem Artikel die Bezeichnung *„Logik und Grundlagenforschung"* für das von Scholz umrissene Gebiet verwendet, ohne dass wir anfangs den Charakter und Umfang dieses Gebiets genauer eingrenzen; ein genaueres Verständnis davon, was Scholz unter diesem Begriffskomplex verstand, ist eines der Ziele dieses Artikels.

Im Jahre 1938 setzte sich Scholz unter Verwendung regimekonformen Vokabulars auf der politischen Ebene dafür ein, daß dieses von ihm benannte Gebiet eine prominente Stellung in der deutschen Wissenschaftslandschaft einnehmen sollte. In einer Denkschrift an den nationalsozialistischen Reichsminister für Wissenschaft (Bernhard Rust, 1883–1945) beschreibt Scholz das Gebiet und beklagt, daß diese „Schöpfung des deutschen Geistes" sich in Polen und den Vereinigten Staaten durchgesetzt hat, aber „in Deutschland ... trotz Hilbert ganz unverhältnismässig zurückgeblieben" sei:[6]

---

[4] Vgl. Peckhaus (1987).
[5] Die Bezeichnung des Scholz'schen Lehrstuhls von 1938 bis 1943 war *Philosophie der Mathematik und Naturwissenschaften (mit besonderer Berücksichtigung der neuen mathematischen Logik und Grundlagenforschung)* (Peckhaus, 1987, 2007). Ab 1943 heißt der Scholz'sche Lehrstuhl *Mathematische Logik und Grundlagenforschung* und im Jahre 1950 wird das Logistische Seminar in das heute noch existierende *Institut für mathematische Logik und Grundlagenforschung* umbenannt. Peckhaus (2007, S. 104) konstatiert, daß „mit dieser Institutsgründung ... der Institutionalisierungsprozess der mathematischen Logik in Deutschland abgeschlossen [war]". Vgl. auch (Pohlers, 2022).
[6] Zitate aus (Scholz, 1938a, S. 1f). Der Denkschrift liegt ein Unterstützungsschreiben vom „Führer der polnischen Logistiker Herrn Professor Dr. Jan Łukasiewicz–Warschau" (datiert auf den 8. Februar 1936) bei; außerdem verweist Scholz auf den deutschen Botschafter in Warschau, Hans-Adolf von Moltke (1884–1943), der bestätigen könne, dass Łukasiewicz ein „Forscher von Weltruf" sei (S. 7). Diese Denkschrift wurde von Peckhaus

> Es gibt heute eine mathematische Logik und Grundlagenforschung, [die], wie alle positiven Grundwissenschaften nicht-historischer Art, aus der Philosophie hervorgegangen [ist]. Jn den letzten zwanzig Jahren ist sie zu einer selbständigen Wissenschaft geworden, ... die mit anerkannten mathematischen Methoden und mathematisierten Konstruktionsmitteln und Fragestellungen arbeitet und auf dieser Basis ... eine grosse Zukunft hat. (Scholz, 1938a, S. 1)

Scholz' Bemühungen um die Markenbildung der *Logik und Grundlagenforschung* waren erfolgreich und der Begriff findet sich innerhalb und außerhalb Deutschlands in verschiedenen Namensgebungen wieder: in der 1950 von Jürgen von Kempski (1910–1998) und Arnold Schmidt gegründeten Zeitschrift *Archiv für mathematische Logik und Grundlagenforschung*,[7] im 1952 gegründeten *Instituut voor Grondslagenonderzoek* an der *Universiteit van Amsterdam*,[8] in der 1955 von Günter Asser (1926–2015) und Karl Schröter (1905–1977) in der DDR gegründeten *Zeitschrift für Mathematische Logik und Grundlagen der Mathematik*,[9] der 1962 errichtete *Lehrstuhl für Logik und Grundlagenforschung* an der Rheinischen Friedrich-Wilhelms-Universität Bonn[10] und das dazu gehörende bis 2014 existierende Promotionsfach *Logik und Grundlagenforschung* in Bonn.[11]

## 2 Die Neuordnung der globalen Wissenschaftswelt

In den Jahren nach dem zweiten Weltkrieg ordnete sich die globale Wissenschaftswelt neu: am 16. November 1945 wurde die *United Nations Educa-*

---

im Behmann-Nachlaß (Staatsbibliothek zu Berlin, Preußischer Kulturbesitz, Nachlaß 335) eingesehen, konnte aber auf Nachfrage bei der Staatsbibliothek derzeit nicht im Nachlaß gefunden werden (persönliche Nachricht, 25. Februar 2022). Peckhaus stellte ein Digitalisat zur Verfügung, bei dem allerdings die letzte Seite und die Datierung fehlen. Peckhaus (2018) datiert die Denkschrift auf den 15. Januar 1938. Scholz verfaßte im Jahre 1938 mehrere Denkschriften, die er an den Reichsminister schickte; z.B. (Scholz, 1938b).

[7] Im Geleitwort zur ersten Ausgabe des ersten Bandes (1950) schreiben die Herausgeber: „Wenn mit dem ‚Archiv für mathematische Logik und Grundlagenforschung' die im Titel bezeichnete Disziplin auch im deutschen Sprachgebiet ein eigenes Organ erhält, so bedarf das heute keiner Rechtfertigung mehr." Die Zeitschrift wurde im Jahre 1988 in *Archive for Mathematical Logic* umbenannt.

[8] Vgl. (van Ulsen, 2000, S. 35f). Das Institut wird heute als Vorgängerinstitution des *Institute for Logic, Language and Computation* (ILLC) an der *Universiteit van Amsterdam* verstanden.

[9] Im Vorwort zum ersten Heft des ersten Bandes (1955) schreiben Asser und Schröter: „Die mathematische Logik, die bekanntlich aus der sog. formalen Logik entstanden ist, hat sich inzwischen zu einer selbständigen mathematischen Disziplin entwickelt. ... [E]s sind die mathematischen Untersuchungen, die dieser Disziplin heute das Gepräge geben. (S. 1)". Die Zeitschrift wurde 1993 in *Mathematical Logic Quarterly* umbenannt und erscheint seit 2011 unter der wissenschaftlichen Schirmherrschaft der DVMLG.

[10] Vgl. (Brendel und Stuhlmann-Laeisz, 2022).

[11] Vgl. Neufassung der Promotionsordnung der Philosophischen Fakultät der Rheinischen Friedrich-Wilhelms-Universität Bonn vom 04. Juni 2010; § 5.

*tional, Scientific and Cultural Organization* (UNESCO) gegründet und unterzeichnete am 16. Dezember 1946 eine Kooperationsvereinbarung mit dem seit 1931 existierenden *International Council of Scientific Unions* (ICSU).[12] Die Disziplinen und ihre Vertretungen positionierten sich, um innerhalb diesen neugegründeten Institutionen für ihr jeweiliges Gebiet zu sprechen. Dafür wurden internationale Vereinigungen gegründet, die diesen Vertretungsanspruch wahrnehmen sollten.

Verschiedene Persönlichkeiten aus dem Umfeld der *Logik und Grundlagenforschung* versuchten, innerhalb dieses institutionellen Gefüges einen Ort für ihr Forschungsgebiet zu etablieren. Die *Association for Symbolic Logic* (ASL) existierte bereits seit 1936,[13] die *Société Internationale de Logique et de Philosophie des Sciences* (SILPS) wurde am 10. November 1946 unter Beteiligung von Paul Bernays (1888–1977), Evert Willem Beth (1908–1964), Józef Maria Bocheński (1902–1995) und Gonseth in Bern gegründet,[14] im Jahre 1947 gründete Stanisłas Dockx (1901–1985) die *Académie Internationale de Philosophie des Sciences* (AIPS), mit SILPS als Gründungsmitglied folgte die *Fédération Internationale des Sociétés de Philosophie* (FISP) im Jahre 1948 und dann, mit FISP als Mitglied, das *Conseil International de Philosophie et des Sciences Humaines* (CIPSH) auf Empfehlung des ersten UNESCO-Direktors Sir Julian Huxley (1887–1975) im Jahre 1949. Auf der mathematischen Seite wurde die *International Mathematical Union* (IMU) im Jahre 1950 wiedergegründet (nach der ursprünglichen Gründung im Jahre 1920 und einer Auflösung im Jahre 1932).[15]

Die genannten Logiker und Wissenschaftsphilosophen hatten Bedenken gegen eine einseitige Unterordnung des Forschungsgebiets in die Mathematik oder die Philosophie:

> The prominent Dutch and European logician-philosopher Beth held that logic and philosophy of science were closely connected; it was bad to separate them. Moreover, only a coalition of the two could achieve independence and university-wide influence. If logic remains dependent on others, e.g. the mathematicians, then it will be in a marginalized and disdained position, and contacts with philosophy of science and other sciences will be cut off. This was the reason for many people to avoid organizing within the International Mathematical Union (IMU). ... But on the other hand, nobody wanted to be dependent on the philosophers either ... In the view of the well-known historian of logic Bochenski [Brief von Bocheński an Quine vom 15.

---

[12]Vgl. (Petitjean, 2006; Greenaway, 1996). ICSU fusionierte im Jahre 2018 mit dem *International Social Sciences Council* (ISSC) zum *International Science Council* (ISC).

[13]Die Vorsitzenden in den für diese Diskussion relevanten Jahren waren Alfred Tarski (1901–1983), Ernest Nagel (1901–1985), J. Barkley Rosser (1907–1989) und Willard Van Orman Quine (1908–2000).

[14]Vgl. (van Ulsen, 2017, S. 71).

[15]Vgl. (Lehto, 1998).

Mai 1953]: „We have to fight against both mathematicians and philosophers for its [formal logic's] recognition." (van Ulsen, 2022)[16]

Die Alternative war die Gründung einer eigenen globalen Vereinigung für das umrissene Gebiet. Im Jahre 1949 gründeten Dockx und Gonseth die *Union Internationale de Philosophie des Science*, welche das Gebiet in ICSU vertreten sollte und als Dachorganisation der anderen Gesellschaften intendiert war. Gonseth warb unter den Vertretern des Gebiets in den verschiedenen Ländern um die Gründung von nationalen Sektionen (später: Nationalkomitees), die die jeweiligen Nationen vertreten konnten.

Gonseth hatte Scholz am 21. April 1950 um die Gründung einer „deutschen Sektion der Internationalen Gesellschaft für Philosophie der Wissenschaften und Logik" (SILPS) gebeten;[17] Scholz war zunächst skeptisch über die Einbettung der Logik in die Wissenschaftsphilosophie und hätte eine direkte Vertretung der Logik ohne Verbindung zur Wissenschaftsphilosophie vorgezogen.[18] Daher erteilte Scholz Gonseth zunächst eine Absage,[19] wurde dann aber von Beth und Stephen Kleene (1909–1994), die ihn im Sommer in Münster besuchen, anderweitig überzeugt.[20] Am 3. August 1950 schrieb Scholz acht deutsche Logiker an:

> Jch bringe den oben genannten Herren zur Kenntnis, dass ich dringend gebeten worden bin, eine deutsche Gesellschaft für mathematische Logik und Grundlagenforschung ins Leben zu rufen, damit die durch sie repräsentierte deutsche Forschung auf eine angemessene Art beteiligt werden kann an den Mitteln, die die Unesco auswerfen will für die Philosophy of Science. ... Es kommt darauf an, dass für die mathematische Logik und Grundlagenforschung der ihr zukommende Raum rechtzeitig gesichert wird. ... Um das Gewicht unserer Forschung zu erhöhen, sollen so bald als möglich möglichst viele zusätzliche nationale Verbände derselben Art ins Leben gerufen werden. Jch bitte die genannten Herren im Jnteresse der guten Sache, mir so bald als möglich unverbindlich mitzuteilen, ob sie bereit sein würden, ihren Namen für die geplante Schöpfung zur Verfügung zu stellen.[21]

---

[16]Vgl. auch den Brief von Beth an Rosser vom 21. Dezember 1952: „As you know, I do not think that this [association with IMU] would be a wise decision. In most countries, mathematicians are hardly interested in symbolic logic or in the philosophy of science, so the influence of the logicians in the Mathematics Union would be very small." (Zitiert nach van Ulsen, 2000, S. 29.)

[17]Brief von Gonseth an Scholz v. 21. April 1950 (Scholz-Nachlaß 111,089).

[18]Brief von Scholz an Alonzo Church (1903–1995) v. 9. Mai 1950 (Scholz-Nachlaß).

[19]Brief von Gonseth an Scholz v. 10. Mai 1950 (Scholz-Nachlaß 111,090).

[20]Briefe von Kleene an Scholz v. 17. Mai, 30. Mai und 7. Juni 1950 (Scholz-Nachlaß 112,005, 112,006 und 112,007); Brief von Scholz an Rosser v. 13. Juli 1950 (Scholz-Nachlaß); in diesem Brief schlägt Scholz vor, dass die UIPS eine Unterorganisation der ASL werden sollte.

[21]Brief von Scholz an Wilhelm Ackermann (1896–1962), Heinrich Behmann (1891–

Dieser Brief ist die erste Erwähnung der Gesellschaft und somit ist der in ihm genannte Name die erste Version des Namens der späteren DVMLG. Bereits am 29. August 1950 kann Scholz Gonseth mitteilen, daß alle acht Herren „grundsätzlich bereit [sind], einer solchen Vereinigung beizutreten".[22] Allerdings bleibt die Angelegenheit danach liegen, weil Scholz Gonseth nicht vollständig vertraut[23] und erst nach einem Austausch des Vorstands der UIPS setzte Scholz seine Vorbereitungen der Gründung der DVMLG zur Gründung einer deutschen Sektion in der UIPS fort.[24] Er delegierte Aufgaben an Britzelmayr, Lorenzen, Schmidt und Schröter, so daß dann die DVMLG (zum ersten Male) gegründet werden konnte.[25] Der erste Satzungsentwurf aus dem Jahre 1954 enthält den jetzigen Namen *Deutsche Vereinigung für mathematische Logik und für Grundlagenforschung der exakten Wissenschaften*.[26] Diese Vereinigung vertrat dann das *nationale Mitglied Bundesrepublik Deutschland* in der UIPS-Nachfolgeorganisation DLMPS

---

1970), Hermes, von Kempski, Paul Lorenzen (1915–1994), Arnold Schmidt, Schröter und Kurt Schütte (1909–1998) v. 3. August 1950 (Scholz-Nachlaß; Unterstreichung im Original).

[22] Brief von Scholz an Gonseth v. 29. August 1950 (Scholz-Nachlaß). Vgl. auch Brief von Lorenzen an Scholz v. 4. August 1950 (Scholz-Nachlaß 112,037).

[23] Brief von Scholz an Arnold Schmidt v. 14. Oktober 1952 (Scholz-Nachlaß): „[Die] Zusammenfassung der Repräsentanten der mathematischen Logik und Grundlagenforschung im deutschen Raum mit dem Ziel einer Eingliederung in die Unesco ... ist in der letzten Augustwoche ... in Paris besprochen worden. ... [V]or rund 2 Jahren [hatte ich] die Sache schon fast auf die Beine gestellt ... Jch habe sie aber damals liegengelassen, weil die Korrespondenz mit Herrn Gonseth so undurchsichtig war, dass ich mich nicht entschliessen konnte, unsere Sache in seine Hände zu legen."

[24] Hodges (2015, S. 12) beschreibt die Vorgänge als „a coup in [UIPS]"; Feferman und Feferman (2004, S. 250) nennen sie einen „putsch by Beth and his friends". Vgl. hierzu einen Brief von Robert Feys (1889–1961) an Rosser, September 1952, zitiert nach van Ulsen (2022): „With the elimination of prof. Gonseth, Dockx and Bayer from the committee, the fundamental difficulties to the adhesion of the ASL to [UIPS] are removed. ... These three colleagues ... were interested in rather literary forms of 'Philosophy of Science'."

In diesem Zusammenhang ist auch der direkte, aber nicht namentliche Angriff auf Gonseth in Quines *ASL Presidential Address* vom 4. September 1953 zu sehen: „The network of organised academia in Europe, the [UIPS] and ICSU and CIPSH and FISP and SILPS and their bewildering kin, are not without unscrupulous men greedy for power and influence. ... Such men ... have the strength of many; for science claims none of their energy and conscience none of their resolve." (van Ulsen, 2000, S. 28)

Die personellen Änderungen an der Spitze der UIPS erlaubten den Beitritt der ASL zur UIPS und die Fusion der UIPS mit der *Union Internationale de Histoire des Sciences* (UIHS). In der neuen, fusionierten Institution, der *International Union of History and Philosophy of Science* (IUHPS) hieß die Nachfolgeorganisation der UIPS nun *Division for Logic, Methodology and Philosophy of Science* (DLMPS) mit expliziter und prominenter Erwähnung der Logik im Namen der Institution. Vgl. auch (van Ulsen, 2022) und (Feferman und Feferman, 2004, S. 248–253).

[25] Vgl. Brief von Scholz an Gonseth v. 19. Juni 1952 (Scholz-Nachlaß) und Brief von Scholz an Jean-Louis Destouches (1909–1980) v. 9. August 1952 (Scholz-Nachlaß).

[26] DVMLG-Archiv A2. Vgl. auch (Löwe und Sarikaya, 2022).

(s. Anm. 24). In den schlecht dokumentierten frühen Jahren der DLMPS bis 1960 war Arnold Schmidt einer ihrer Vorsitzenden.

## 3  Die exakten Wissenschaften

Zwischen Scholz' Brief vom 3. August 1950 (s. Anm. 21) und der Gründung der DVMLG war der Begriff „Grundlagenforschung" durch die zusätzliche Spezifikation „der exakten Wissenschaften" eingeengt worden. Es ist interessant festzustellen und sicherlich kein Zufall, dass die niederländische Schwestergesellschaft der DVMLG, die *Nederlandse Vereniging voor Logica en Wijsbegeerte der Exacte Wetenschappen* (VvL) ebenfalls diesen Begriff im Namen trägt.[27]

Im wissenschaftsphilosophischen Diskurs der 1940er und 1950er Jahre hatten die Begriffe *deduktive Wissenschaften* (Tarski, 1994) und *Formalwissenschaften* (Carnap, 1935) eine deutlich umrissene Bedeutung und bezeichneten „Logik, einschließlich der Mathematik" (Carnap, 1935, S. 30). Der Begriff der *exakten Wissenschaften* ist weniger klar definiert. In der antiken und mittelalterlichen Tradition werden als *exakte mathematische Wissenschaften* die sogenannten *scientiae mediae* (oft Optik, Astronomie und Statik) bezeichnet:

> [T]he exact mathematical sciences fell somewhere between natural philosophy and pure mathematics, perhaps closer to the latter than the former. But the exact sciences belong neither wholly to natural philosophy nor to mathematics but are relevant to both. Because they were viewed as lying between the two disciplines, the exact sciences came to be known as middle sciences (scientiae mediae) during the Middle Ages. (Grant, 2007, S. 43)

Grant (2007, Kapitel 10, insbesondere S. 319–322) beschreibt, wie sich die Bedeutung des Begriffs „exakte Wissenschaften" in der frühen Neuzeit verschiebt und die Physik mit einschliesst. Der Begriff wird regelmäßig in der heutigen Wissenschaftswelt verwendet, sein Umfang divergiert allerdings erheblich: in manchen Kontexten sind lediglich Mathematik, Informatik und Teile der Physik gemeint, manchmal auch der Rest der Physik und die Chemie und manchmal sind zusätzlich mathematische Teile der Technikwissenschaften Teil des Umfangs.[28]

---

[27]Die VvL war am 15. November 1947 von Evert Willem Beth (1908–1964), Arend Heyting (1898–1980) und A. G. M. van Melsen (1912–1994) gegründet worden. Vgl. (van Ulsen, 2017, S. 72f) und (van Benthem, 2003).

[28]Z.B. hatte die *Nederlandse Organisatie voor Wetenschappelijk Onderzoek* (NWO) bis vor wenigen Jahren eine Organisationseinheit *Exacte Wetenschappen*, welches aus Astronomie, Informatik und Mathematik bestand; die *Raymond & Beverly Sackler Faculty of Exact Sciences* an der *Tel Aviv University* umfasst Chemie, Informatik, Geowissenschaften, Mathematik, Physik und Astronomie; der Studiengang *Bacharelado em*

In Scholz' hochschulpolitischen Denkschriften findet sich der Begriff der exakten Wissenschaften nicht: er erwähnt zwar eine „generalisierte Grundlagenforschung", beschreibt aber im Detail nur die „auf die Mathematik spezialisierte Grundlagenforschung" (Scholz, 1938a, S. 5). In einem Schreiben an Karl Friedrich Schmidt, den damaligen Direktor des Mathematischen Instituts in Münster, diskutiert Scholz die Möglichkeit einer Grundlagenforschung der Physik oder Biologie. Er verwirft ersteres, weil „der einzige, der das wirklich vermag ... ist ... Carl Friedrich v. Weizsäcker, [der] ... sich von der Atomphysik nicht trennen [wird]" und letzteres, weil in der Biologie „alles, was sich ... als philosophische ... Durchdringung bezeichnet, ... unscharf und ... verschwommen" sei.[29] Diese Einstellung entspricht der von Tarski, der im Vorwort zur ersten amerikanischen Ausgabe (1941) seines Buchs *Introduction to Logic and to the Methodology of the Deductive Sciences* schrieb:

> I am inclined to doubt whether any special 'logic of empirical sciences' ... exists at all. ... [T]he methodology of empirical sciences constitutes an important domain of scientific research. ... [U]p to the present, logical concepts and methods have not found any specific or fertile applications in this domain. And it is certainly possible that this state of affairs is not simply a reflection of the present stage of methodological research. Perhaps it arises from the circumstance that ... an empirical science may have to be considered not simply as a scientific theory ... but rather as a complex, consisting partly of [the statements of a theory] and partly of human activities. ... [T]he methodology of these sciences can hardly boast of ... definitive achievements—despite the great efforts that have been made. (Tarski, 1994, S. xii–xiii)

Weder Scholz noch Tarski schränken den Wirkungsbereich der *Grundlagenforschung* grundsätzlich ein; sie behalten sich die Option einer „generalisierten Grundlagenforschung" vor, bestanden aber auf einer Grundlagenforschung die „mit anerkannten mathematischen Methoden und mathematisierten Konstruktionsmitteln und Fragestellungen arbeitet" (Scholz, 1938a, S. 1). Für Scholz war es eine kontingente Tatsache der zeitgenössischen Forschungslage in der Grundlagenforschung der empirischen Wissenschaften, daß es für diese keine mathematisch-exakte Grundlagenforschung gab; somit mußte sich die Grundlagenforschung auf diejenigen Disziplinen einschränken, in denen eine Grundlagenforschung „mit anerkannten mathematischen Methoden" existierte.

---

*Ciências Exatas* an der *Universidade Federal de Juiz de Fora* umfasst die Fächer *Ciência da Computação, Engenharia Computacional, Engenharia Elétrica, Engenharia Mecânica Estatística, Física, Matemática* und *Química*.

[29] Brief von Scholz an Friedrich Karl Schmidt v. 17. Dezember 1950 (Scholz-Nachlaß), S. 6 & 7.

Es ist also davon auszugehen, daß durch die Spezifikation „der exakten Wissenschaften" im Namen der DVMLG weniger eine konkrete Einschränkung auf bestimmte zu betrachtende Wissenschaften als vielmehr eine methodische Einschränkung der Grundlagenforschung auf die mathematisch-exakten Methoden intendierte. Dies entspricht der Verwendung des Wortes „exakt" in der Denkschrift von 1938, in der Scholz als neuen Titel für seinen Lehrstuhl „Professur für Logik und exakte Philosophie der Mathematik und Naturwissenschaften" vorschlägt.[30]

## 4 Ist Grundlagenforschung Philosophie?

In den Abschnitten 2 und 3 wurde der Begriff *Logik und Grundlagenforschung* für die von Scholz intendierte Forschungsgemeinschaft verwendet, ohne genauer zu definieren, um welche Gemeinschaft es sich handelt. Eingangs war erwähnt worden, dass dieser Begriff bisweilen als Hendiadyoin verstanden wird und dieser Forschungszweig als „mathematische Subdisziplin" (Peckhaus, 2007, S. 103) bezeichnet wird.

Die Frage nach dem Verhältnis der Grundlagenforschung zur Philosophie stellte sich nicht erst in den 1950er Jahren. Bereits beim gescheiterten Versuch der von Gerhard Hessenberg (1874–1925) betriebenen Gründung der *Zeitschrift für die Grundlagen der gesamten Mathematik* im Jahre 1908 gerät der Versuch, der „streng mathematischen Richtung [der Grundlagenforschung] ein Organ [zu] schaffen",[31] in Kritik. Alexander Rüstow vom Teubner-Verlag antwortet auf Hessenbergs Vorschlag und beklagt den Ton,

> aus welchem dauernd ein etwas reizbares odi profanum vulgus et arceo als Unterton mitklingt, wobei unter dem fernzuhaltenden Pöbel aber schlechthin jeder Nicht-Fach-Mathematiker gemeint war. ... [E]s ist doch klar, dass man für die wichtigsten Probleme der Grundlagen der Mathematik ebenso sehr Philosoph wie Mathematiker sein muss.[32]

Jahrzehnte später vertrat Heinrich Scholz eine Position, die der von Peckhaus (2007) beschriebenen Position Hessenbergs sehr ähnlich war: er

---

[30]Vgl. (Scholz, 1938a, S. 7). Der Schweizer Wissenschaftssoziologie Emil J. Walter (1937, S. 2) verwendet ebenfalls den Begriff der „exakten Wissenschaften" bei seiner Beschreibung „der gegenwärtig wohl dringlichsten wissenschaftstheoretischen Aufgabe, die methodischen Fortschritte der exakten Wissenschaften anderen, in ihrer inneren Entwicklung zurückgebliebenen Realwissenschaften zugänglich zu machen." Er führt aus: „[E]s steht [zu erwarten], die Uebertragung gewisser, in den exakten Wissenschaften gewonnener Resultate auf andere, nicht so ausgebaute Zweige der Einheitswissenschaft werde das wissenschaftliche Streben in Zukunft noch mehr als bisher befruchten können (S. 20)." In seinen wissenschaftsklassifikatorischen Schriften verwendet Walter (1943) ausschließlich die Carnapsche Dichotomie der Formal- und Realwissenschaften und nicht den Begriff der „exakten Wissenschaften".

[31]Brief von Hessenberg an den Teubner-Verlag v. 29. April 1908; zitiert nach Peckhaus (2007, S. 110).

[32]Brief von Rüstow an Hessenberg v. 29. Oktober 1908, zitiert nach Peckhaus (2007, S. 111).

sprach einer methodisch exakten Philosophie eine wichtige Rolle in der Grundlagenforschung zu, forderte aber einen klaren Primat der mathematischen Methoden und war gegenüber anderen Strömungen der Philosophie äußerst skeptisch. Für Scholz sind Logik und Grundlagenforschung nicht identisch und beide Gebiete sind zumindest traditionell in der Philosophie zu verorten:

> Logik und Grundlagenforschung sind seit Aristoteles zwei anerkannte philosophische Disziplinen im abenländischen [sic!] Raum. Es liegt nicht an mir, sondern an einem Gang der Dinge, den niemand aufzuhalten vermocht hat, dass die Logik mit der auf ihr fussenden Grundlagenforschung, insbesondere mit der hoch entwickelten Wissenschaftslehre der deduktiven Wissenschaften, heute nur noch durch mathematische Methoden und Fragestellungen beherrscht und gefördert werden kann.[33]

In seinen hochschulpolitischen Denkschriften betont er den philosophischen Charakter von *Logik und Grundlagenforschung*:

> [P]hilosophisch ist diese neue Logik und Grundlagenforschung ... in dem Sinne, dass sie primär auf eine Tieferlegung und Klärung der Fundamente gerichtet ist und dass sie wenigstens bis jetzt noch keine Methoden hervorgerufen hat, für welche gezeigt werden konnte, dass mit Hilfe dieser Methoden innermathematische Probleme gelöst werden können, die sich vor der Entdeckung dieser Methoden einer Lösung entzogen haben (Scholz, 1938a, S. 3).

Gleichzeitig meinte Scholz, dass er unter den Ordinarien der Philosophie in Deutschland keine Verbündeten für *Logik und Grundlagenforschung* finden kann und dass man dieses Gebiet gegen negative Einflüsse der herkömmlichen Philosophie beschützen mußte:

> Jch habe den Kreis so eng gehalten, um sicher zu sein, dass wir gegen den Einbruch der Dilettanten und Schwätzer in jedem Fall gesichert sind. Jch habe Grund hinzuzufügen, dass die ernsten Befürchtungen, die ich schon im Januar geltend gemacht habe auf eine mir ganz und gar nicht erwünschte Art bestätigt worden sind durch die seltsamsten Dinge, die sich im Namen der mathematischen Logik inzwischen in München zugetragen haben. Mit diesen Leuten kann und werde ich in keinem Falle zusammengehen; denn sie haben einen anderen Geist als wir.[34]

---

[33] Brief von Scholz an Friedrich Karl Schmidt v. 17. Dezember 1950 (Scholz-Nachlaß; Unterstreichung im Original). Vgl. auch (Scholz, 1938a, S. 6): „Die eine [Brücke in die Zukunft] muß von der Philosophie ausgehen; denn es ist unrühmlich für uns, wenn wir noch länger zulassen, daß die Logik und die wissenschaftsverbundene Grundlagenforschung, die seit Aristoteles zwei unbestritten philosophische Grundlagendisziplinen sind, nur so weit der Mühe wert sein sollen, wie sie im klassischen Sinne behandelt werden."
[34] Brief von Scholz an Gonseth v. 29. August 1950 (Scholz-Nachlaß).

Dieser „andere Geist" ist eine Bezugnahme auf den Psychologismusstreit; Scholz kritisiert eine Form der Philosophie, welche keine exakten, sondern lediglich historische und psychologische Methoden verwendet.[35] Es ist auffällig (aber nicht einfach zu interpretieren), dass Scholz in seiner gesamten (vollständig deutschsprachigen) Korrespondenz mit Gonseth, Rosser und den Gründungsmitgliedern der DVMLG durchgehend den englischen Begriff *„philosophy of science"* verwendet.[36]

Methodisch eng an der Mathematik orientiert und dennoch von philosophischem Charakter berührt also die *Logik und Grundlagenforschung* beide Disziplinen. Allerdings ist die Wissenschaftswelt stark disziplinär geordnet, und ein interdisziplinäres Forschungsgebiet kann zwischen die Stühle fallen, wenn es sich nicht an einer der beiden relevanten etablierten Disziplinen orientiert. Konfrontiert mit der Frage, welche der beiden Disziplinen man wählen sollte, wenn man gezwungen wäre, ist Scholz' Antwort klar und eindeutig: die Mathematik.

> Es sollte auf keine Art verantwortet werden können, dass man, aus welchem Grunde auch immer, dies opfert, um den alten elenden Schlendrian wieder hereinzulassen, Hierfür wird man in jedem Falle verantwortlich sein, wenn ein Philosoph im alten Sinne für meinen Lehrstuhl vorgeschlagen wird. Das, was zu fordern ist, kann nur von einem gelernten Mathematiker geleistet werden.
>
> [Es] muss mit der Möglichkeit gerechnet werden, dass der bis jetzt bestehende Zusammenhang mit der Philosophie in Frage gestellt wird. ... Es scheint mir nicht unmöglich zu sein, ... zum Ausdruck zu bringen, wie sehr es im Jnteresse eines vorschreitenden, nicht nur im Historischen oder Psychologischen stecken bleibenden Philosophierens erwünscht ist ..., dass eine Prüfung in mathematischer Logik und Grundlagenforschung als philosophische Prüfung anerkannt wird. ... Wenn dies nicht gelingt, so sollte die mathematische Logik und Grundlagenforschung in die Mathematik ... eingebaut werden. ...
>
> Ich kann natürlich nicht bestreiten, daß in diesem Falle formal eine philosophische Professur für die Fakultät verloren geht. Aber auch nur formal, in keinem Fall in einem tiefer liegenden Sinne; denn ... Logik bleibt Logik, Grundlagenforschung bleibt Grundlagenforschung. Ob man bereit ist, sie auch noch auf der Stufe der heute erreichbaren Standfestigkeit als philosophische Disziplinen anzuerkennen, oder ob man umgekehrt gewillt ist, ihnen um dieser Zuverlässigkeit willen die philosophische Würde abzusprechen, ist eine Entscheidungsfrage, die

---

[35] Vgl. Rath (1994), Galliker (2016, Abschnitt 5.2) und Kusch (2020). Vgl. auch Feys' Vorwurf, dass Gonseth, Dockx und Bayer an einer „literary form of 'Philosophy of Science'" interessiert seien (Anm. 24).

[36] Brief von Scholz an Rosser v. 13. Juli 1950, Brief von Scholz an Ackermann, Behmann, Hermes, von Kempski, Lorenzen, Arnold Schmidt, Schröter, Schütte v. 3. August 1950 und Brief von Scholz an Gonseth v. 29. August 1950 (Scholz-Nachlaß).

an ihrer Substanz und an ihrem Charakter in gar keinem Fall etwas ändern wird.³⁷

Innerhalb der DVMLG entwickelt sich die Frage nach dem Verhältnis zwischen Logik und Philosophie zu einem Streit in den Jahren 1978 bis 1980: nach einer Diskussion auf der Mitgliederversammlung am 3. Oktober 1978 in Aachen wurden Peter Janich und Christian Thiel aufgefordert, ein Memorandum zu den „wissenschaftlichen Aktivitäten der Gesellschaft auf dem Gebiet der Grundlagenforschung der exakten Wissenschaften sowie [der Intensivierung der] Mitgliedergewinnung unter Wissenschaftlern dieser Forschungsrichtung" zu verfassen.³⁸ Dieses Memorandum wurde im November 1978 mit dem Protokoll der Mitgliederversammlung an die Mitglieder verschickt:

> Die Deutsche Vereinigung für mathematische Logik und für Grundlagenforschung der exakten Wissenschaften (DVMLG) hat sich selbst in ihrer Satzung die Aufgabe gestellt, neben der mathematischen Logik auch die ‚Grundlagenforschung der exakten Wissenschaften' zu fördern. ... [D]ie Zusammensetzung der Mitgliedschaft [hat sich] weitgehend zu Gunsten der Vertreter der mathematischen Logik verschoben. Wir möchten anregen, die philosophisch-wissenschaftstheoretische Fraktion wieder zu beleben.³⁹

Das Memorandum löste eine vehemente Debatte aus. Walter Felscher verfasste im Februar 1979 eine Replik, die ebenfalls an alle Mitglieder versandt wurde:

> [D]ie DVMLG [ist] (ebenso wie übrigens die ASL) keine Vereinigung zur Beförderung der Philosophie—handele es sich nun um die diversen Philosophien der Mathematik ... oder um die verschiedenen, neuerdings als Wissenschaftstheorien bezeichneten philosophischen Bemühungen. ... Es besteht ... gar kein Grund, die DVMLG zum Orte solcher Aktivitäten zu machen und, wie es im Memorandum heißt, ihren *institutionellen Rahmen* für solche, der Mathematischen Logik gänzlich fremdartigen Beschäftigungen zu nutzen. ... Im Übrigen würde eine solche außerwissenschaftliche Aktivität die Vereinigung und die Mathematische Logik überhaupt unter den Kollegen der Mathematik und der Naturwissenschaften erneut in Verruf bringen, nachdem sich während der letzten 30 Jahre die Mathematische Logik gerade von dem Odeur philosophisch-ideologischer Doktrinen

---

³⁷Brief von Scholz an Friedrich Karl Schmidt v. 17. Dezember 1950 (Scholz-Nachlaß; Unterstreichung im Original), S. 4–6.
³⁸DVMLG-Archiv A84.
³⁹Memorandum von Peter Janich und Christian Thiel v. 21. November 1978 (DVMLG-Archiv A87).

befreit und als eine seriöse mathematische Disziplin Verständnis gefunden hat.[40]

Der damalige Vorsitzende, Gert H. Müller versuchte, in einem Rundschreiben und einer informellen Mitgliederbesprechung in Oberwolfach am 26. April 1979, den Streit zu schlichten.[41] Im Protokoll der nächsten Mitgliederversammlung am 24. August 1979 in Hannover (bei der Felscher anwesend war) gibt es allerdings keine Anzeichen größerer Differenzen: es wird angeregt,

> dass der Vorstand Tagungen veranstalten möge, die das gesamte Spektrum des Vereins abdecken. Herr Müller verweist auf seinen Vorschlag, auch in Oberwolfach andere Themen von Zeit zu Zeit miteinzubeziehen. Herr Thiel berichtet, dass von ihm selbst und den Herren W. Oberschelp und M. Richter demnächst eine Tagung vorbereitet wird. Derartige Initiativen werden von der Versammlung begrüßt.[42]

Im folgenden Jahr wurde Christian Thiel in den Vorstand der DVMLG gewählt;[43] in den folgenden Jahren sollte die Frage nach dem Verhältnis zwischen Logik und Philosophie durch die Frage nach dem Verhältnis zwischen Logik und Informatik verdrängt werden, die einen Streit mit größeren Auswirkungen für den Verein auslöste.

## Literaturverzeichnis

Brendel, E. und Stuhlmann-Laeisz, R. (2022). Geschichte des Lehrstuhls für Logik und Grundlagenforschung an der Rheinischen Friedrich-Wilhelms-Universität Bonn. In Löwe, B. und Sarikaya, D., Herausgeber, *60 Jahre DVMLG*, Band 48 in *Tributes*. College Publications.

Carnap, R. (1935). Formalwissenschaft und Realwissenschaft. *Erkenntnis*, 5(1):30–37.

---

[40] W. Felscher, Note zum Memorandum der Herren Janich und Thiel v. Februar 1979 (DVMLG-Archiv A87).

[41] Vgl. Brief von Gert H. Müller an die Mitglieder der DVMLG v. März 1979 (DVMLG-Archiv A87): „Eine Aufspaltung in zwei Vereinigungen oder durch ‚Fraktionen' erscheint mir angesichts der Mitgliederzahl ca. 120 nicht am Platze. ... Wissenschaftstheorie heute ist von Argumenten math.-log.-Art [sic!] stark durchdrungen—nicht zu reden von Denkweisen der Informatik, so dass diese eher in unserer Vereinigung als sonstwo einen Platz findet." Vgl. auch Brief von Gert H. Müller an die Mitglieder der DVMLG v. Juli 1979 (DVMLG-Archiv A88; Unterstreichung im Original): „Bei der informellen Mitgliederbesprechung in Oberwolfach (26.IV.) brachte Herr Felscher ... eine <u>deutlich</u> ablehnende Haltung gegenüber den von mir ... vorgebrachten Vorschlägen zum Ausdruck."

[42] DVMLG-Archiv A90.

[43] Protokoll der Mitgliederversammlung der DVMLG am 19.9.1980 in Dortmund (DVMLG-Archiv A93).

Feferman, A. B. und Feferman, S. (2004). *Alfred Tarski. Life and Logic.* Cambridge University Press.

Galliker, M. (2016). *Ist die Psychologie eine Wissenschaft? Ihre Krisen und Kontroversen von den Anfängen bis zur Gegenwart.* Springer-Verlag.

Grant, E. (2007). *A History of Natural Philosophy. From the Ancient World to the Nineteenth Century.* Cambridge University Press.

Greenaway, F. (1996). *Science International: A history of the International Council of Scientific Unions.* Cambridge University Press.

Hodges, W. (2015). DLMPS—Tarski's vision and ours. In Schroeder-Heister, P., Heinzmann, G., Hodges, W., und Bour, P. É., Herausgeber, *Logic, Methodology and Philosophy of Science. Logic and Science Facing the New Technologies. Proceedings of the 14th International Congress (Nancy)*, Seiten 9–26. College Publications.

Kusch, M. (2020). Psychologism. In Zalta, E. N., Herausgeber, *The Stanford Encyclopedia of Philosophy*. Spring 2020 Edition.

Lehto, O. (1998). *Mathematics without borders. A history of the International Mathematical Union.* Springer-Verlag.

Löwe, B. (2022). Die Mitgliederentwicklung in der Frühzeit der DVMLG. In Löwe, B. und Sarikaya, D., Herausgeber, *60 Jahre DVMLG*, Band 48 in *Tributes*. College Publications.

Löwe, B. und Sarikaya, D. (2022). Satzungen der DVMLG durch die Jahrzehnte. In Löwe, B. und Sarikaya, D., Herausgeber, *60 Jahre DVMLG*, Band 48 in *Tributes*. College Publications.

Peckhaus, V. (1987). Geschichte des Scholz-Seminars 1936–1942 im Spiegel der Vorlesungsverzeichnisse der Universität Münster. Maschinenschriftliche Aktennotiz, Erlangen, 2. November 1987.

Peckhaus, V. (2007). Die Zeitschrift für die Grundlagen der gesamten Mathematik. Ein gescheitertes Zeitschriftenprojekt aus dem Jahre 1908. *Mathematische Semesterberichte*, 54:103–115.

Peckhaus, V. (2018). Heinrich Scholz. In Zalta, E. N., Herausgeber, *The Stanford Encyclopedia of Philosophy*. Fall 2018 Edition.

Petitjean, P. (2006). The Early Yars of UNESCO-ICSU Partnership. In Petitjean, P., Zharov, V., Glaser, G., Richardson, J., de Padirac, B., und Archibald, G., Herausgeber, *Sixty Years of Sciences at Unesco, 1945-2005*, Seiten 77–78. UNESCO.

Pohlers, W. (2022). Eine kurze Geschichte der Entwicklung der Logik in Münster. In Löwe, B. und Sarikaya, D., Herausgeber, *60 Jahre DVMLG*, Band 48 in *Tributes*. College Publications.

Rath, M. (1994). *Der Psychologismusstreit in der deutschen Philosophie*. Alber.

Scholz, H. (1921). *Religionsphilosophie*. Reuther & Reichard.

Scholz, H. (1938a). Denkschrift über die neue mathematische Logik und Grundlagenforschung. An den Herrn Reichs- und Preussischen Minister für Wissenschaft, Erziehung und Volksbildung; Digitalisat, von Volker Peckhaus zur Verfügung gestellt (vgl. Anmerkung 6).

Scholz, H. (1938b). Denkschrift über die Neugestaltung des Hochschulunterrichts in der Philosophie überhaupt und der mathematischen Logik und Grundlagenforschung im besonderen. An den Herrn Reichs- und Preussischen Minister für Wissenschaft, Erziehung und Volksbildung, 28. Dezember 1938 (Scholz-Nachlaß 102,006).

Tarski, A. (1994). *Introduction to Logic and to the Methodology of the Deductive Sciences*, Band 24 in *Oxford Logic Guides*. Oxford University Press, 4. Auflage.

van Benthem, L. (2003). The Dutch Association for Logic and Philosophy of the Exact Sciences (De Nederlandse Vereniging voor Logica en Wijsbegeerte der Exacte Wetenschappen). Ontsluitingsdocument, Volume 860, Noord-Hollands Archief.

van Ulsen, P. (2000). *E. W. Beth als logicus*. Dissertation, Universiteit van Amsterdam. ILLC Publications DS-2000-04.

van Ulsen, P. (2017). Organisaties en genootschappen. Teilbericht des Forschungsprojekts *E. W. Beth and A. J. Heyting. Their influence and ideas on philosophy, logic and related sciences*.

van Ulsen, P. (2022). The birth pangs of DLMPS. In Löwe, B. und Sarikaya, D., Herausgeber, *60 Jahre DVMLG*, Band 48 in *Tributes*. College Publications.

Walter, E. J. (1937). Logistik, logische Syntax und Mathematik. *Vierteljahrsschrift der Naturforschenden Gesellschaft in Zürich*, 82:2–20.

Walter, E. J. (1943). Einheitswissenschaft. *Vierteljahrsschrift der Naturforschenden Gesellschaft in Zürich*, 88:22–35.

# Satzungen der DVMLG durch die Jahrzehnte

## Benedikt Löwe[1,2,3], Deniz Sarikaya[4]

[1]Institute for Logic, Language and Computation, Universiteit van Amsterdam, Postbus 94242, 1090 GE Amsterdam, The Netherlands

[2]Fachbereich Mathematik, Universität Hamburg, Bundesstraße 55, 20146 Hamburg, Deutschland

[3]Churchill College, Lucy Cavendish College, & Department of Pure Mathematics and Mathematical Statistics, University of Cambridge, Storey's Way, Cambridge CB3 0DS, England

[4] Centre for Logic and Philosophy of Science, Vrije Universiteit Brussel, Pleinlaan 2, 1050 Brussel, Belgien

E-Mail: `loewe@math.uni-hamburg.de`, `deniz.sarikaya@vub.be`

## 1 Einleitung

Dieser Artikel enthält die Texte der verschiedenen Satzungen der DVMLG zwischen 1954 und 2022.

Die wesentliche Grundlage des Artikels ist das Archiv der DVMLG. Die Unterlagen bis 1990 aus dem Archiv der DVMLG sind im Jahre 2012 von Peter Koepke und Daniel Witzke konsolidiert, in fünf Ordner (A bis E) gegliedert und teildigitalisiert worden. Die Unterlagen aus späteren Jahren sind bislang nicht archivarisch erschlossen.[1] Die Archivmaterialien befinden sich derzeit an der Universität Hamburg.

Das Archiv enthält auch einige Dokumente aus der Vorgeschichte der DVMLG: die Gründung der DVMLG ist unmittelbar mit der Repräsentation Deutschlands in internationalen Wissenschaftsorganisationen, insbesondere der *Division for Logic, Methodology and Philosophy of Science of the International Union of History and Philosophy of Science* (DLMPS/IUHPS) verbunden. Bereits 1950 hatte sich Ferdinand Gonseth (1890–1975) an Heinrich Scholz (1884–1956) mit der Bitte um Gründung einer deutschen Vertretung der Wissenschaftsphilosophie gewandt. Diese Anfrage führte, vermutlich im Jahre 1954, zur ersten Gründung einer *Deutschen Vereinigung für Mathematische Logik und für Grundlagenforschung der exakten Wissenschaften*.[2]

Diese gleichnamige Vorgängerinstitution der DVMLG stellte in den 1950er Jahren das deutsche Nationalkomitee in der DLMPS, wofür sie ab

---

[1]Die Unterlagen des erschlossenen Archivs bis 1990 werden mit der Archivnummer (Ordnerbuchstabe und laufende Nummer) zitiert; die Unterlagen des nicht erschlossenen Archivs werden mit dem Kürzel NE und der Ordner- und Abschnittsbeschriftung zitiert.

[2]Für eine ausführlichere Schilderung dieser Ereignisse, vgl. B. Löwe, Grundlagenforschung der exakten Wissenschaften: die DVMLG und die Philosophie, in diesem Bande.

1956 einen Jahresbeitrag an die DLMPS zu zahlen hatte.[3] Der damalige Vorsitzende der Vorgängerinstitution, Arnold Schmidt (1902–1967) beantragte die Finanzierung dieser nationalen Mitgliedsbeiträge bei der Deutschen Forschungsgemeinschaft (DFG) und dem Bundesinnenministerium.[4] Die erbetene Finanzierung wurde teilweise in Aussicht gestellt:

> Ich bin grundsätzlich bereit, Ihnen ... den Mitgliedsbeitrag der Deutschen Vereinigung für mathematische Logik und für Grundlagenforschung der exakten Wissenschaften bei der IUHPS zur Verfügung zu stellen, vorausgesetzt, daß Sie durch Vorlage eines Haushaltsplanes Ihrer Vereinigung nachweisen, daß die Vereinigung nicht in der Lage ist, den Mitgliedsbeitrag aus eigenen Mitteln zu bestreiten.[5]

Die sehr informelle Vorgängerinstitution, welche sich bis zu diesem Zeitpunkt nicht mit Einnahmen und Ausgaben befaßt hatte,[6] überlegte nun, ob ein solcher Haushaltsplan die Erhebung von Mitgliedsbeiträgen und damit eine formelle Vereinsstruktur erforderte.[7]

Ein Verein kann durch die Eintragung ins Vereinsregister zu einer juristischen Person (und damit zu einer rechtsfähigen Körperschaft) werden. Das Vereinsregister wird in Deutschland von den Amtsgerichten geführt. Der einzutragende Verein bestimmt durch seine Satzung seine Organisation und die Rechte und Pflichten des Vereins gegenüber seinen Mitgliedern. Beschlossene Satzungsänderungen werden beim Amtsgericht eingereicht, welches überprüft, ob der Beschluß zur Satzungsänderung satzungsgemäß stattgefunden hat.

Die Frage der Gemeinnützigkeit eines Vereins ist separat von der Eintragung ins Vereinsregister: falls der satzungsgemäße Zweck des Vereins durch das Finanzamt als gemeinnützig anerkannt wird, kann der Verein zudem als gemeinnütziger Verein steuerbegünstigt geführt werden.

---

[3] Brief von Hans Freudenthal an Arnold Schmidt v. 1. Januar 1960 (DVMLG-Archiv A5).

[4] Vgl. DVMLG-Archiv A6 bis A10.

[5] Brief von Dr. Petersen an Arnold Schmidt v. 18. April 1961 (DVMLG-Archiv A11).

[6] Vgl. Brief von Arnold Schmidt an Hans Hermes v. 19. Mai 1961 (DVMLG-Archiv A12): "Diese Aufstellung scheint in der Situation unserer Vereinigung—wir haben ja bisher nichts eingenommen—garnicht (sic!) so einfach zu sein."
Es ist zu beachten, daß Freudenthals Rechnung die Jahre 1956 bis 1959 als "payé" markiert (DVMLG-Archiv A5). Es ist unklar, wie diese Beiträge bezahlt wurden; am 19. Mai 1961 schreibt Schmidt an Hermes (DVMLG-Archiv A12): "Daß man uns die Beiträge für IUHPS nicht rückwirkend bewilligen will, halte ich nicht für so sehr schlimm, da Herr Beth meinte, daß diese uns im Falle der von jetzt ab erfolgenden Zahlung erlassen werden würden."

[7] Vgl. Brief von Arnold Schmidt an Hans Hermes v. 5. Juni 1961 (DVMLG-Archiv A14): "[W]ill man überhaupt die DVMLG auf eine solche Basis stellen, bei der dann manches formaler läuft als bisher[?] ... Wenn man aber den Weg der Beitragserhebung beschreiten will: ist dazu nicht vielleicht die Eintragung der Vereinigung in das Vereinsregister notwendig?"

Man entschied sich, diesen formalen Weg zu gehen und gründete den Verein förmlich auf der Gründungssitzung in Marburg am 28. Juli 1962. Es ist hervorzuheben, dass der im Jahre 1962 förmlich gegründete eingetragene Verein von den Mitgliedern als von seiner informellen Vorgängerinstitution separate Entität gesehen wurde: Mitglieder der Vorgängerinstitution wurden nicht automatisch Mitglieder der DVMLG, sondern mussten förmlich um Aufnahme bitten und angenommen werden.[8]

## 2 Chronologischer Überblick

Das Archiv der DVMLG enthält nur wenige Dokumente aus der Zeit vor 1962. Im von Koepke und Witzke erstellten Findbuch finden sich zwei undatierte Satzungsdokumente, beide mit dem Titel *Entwurf. Satzung der deutschen Vereinigung für mathematische Logik und für Grundlagenforschung der exakten Wissenschaften (DVLG)* und beide im Findbuch auf den 18. November 1954 datiert.

Das Dokument A1 ist nahezu identisch mit der im Jahre 1962 an das Amtsgericht übersandten Satzung und erwähnt explizit die Eintragung beim Amtsgericht in Marburg/Lahn (§ 2). Da die Eintragung des Vereins erstmals im Jahre 1961 von Schmidt in Betracht gezogen wurde (vgl. Anm. 7), ist anzunehmen, daß die Datierung im Findbuch inkorrekt ist und es sich bei diesem Dokument um den in der Gründungssitzung am 28. Juli 1962 vorgelegten Satzungsentwurf handelt.

Im Gegensatz dazu hat der Satzungsentwurf in A2 keine Erwähnung der Eintragung ins Vereinsregister und erwähnt einen sechsköpfigen "erweiterten Vorstand" (im Gegensatz zum siebenköpfigen Vorstand in der Satzung von 1962). Das Archiv enthält ein einziges Dokument der Vorgängerinstitution: ein Sitzungsprotokoll vom 18. November 1954, in dem ein sechsköpfiger "erweiterter Vorstand" bestätigt wird.[9] Es ist daher davon auszugehen, daß es sich bei A2 um den im Sitzungsprotokoll erwähnten Satzungsentwurf aus dem Jahre 1954 handelt. Der Text dieser Satzung der Vorgängerinstitution ist in Abschnitt 3 wiedergegeben.

Die DVMLG wird in der Gründungssitzung in Marburg am 28. Juli 1962 in Anwesenheit von Gisbert Hasenjaeger, Hans Hermes, Arnold Schmidt und Kurt Schütte mit schriftlichen Vollmachten von Wilhelm Ackermann, Jürgen von Kempski und Paul Lorenzen gegründet. Kurt Schütte ist der Protokollführer der Gründungssitzung, bei der die Satzung einstimmig angenommen wird und alle sieben Gründungsmitglieder in den Vorstand gewählt werden. Der Verein wird am 13. Dezember 1962 beim Amtsgericht Marburg in das Vereinsregister eingetragen; die dort eingetragene Satzung ist fast

---

[8]Dies wird detailliert beschrieben in B. Löwe, Die frühe Mitgliederentwicklung der DVMLG, in diesem Bande.
[9]DVMLG-Archiv A3.

vollständig identisch mit dem Entwurf in A1.[10] In Abschnitt 4 wird die Gründungssatzung wiedergegeben (mit Abweichungen zwischen A18 und A1 in den Anmerkungen). Die Gründungssatzung hat die folgenden wesentlichen Vorgaben:

(1)[1962] Voraussetzungen für die Mitgliedschaft sind Wohnsitz in Deutschland und eigene Publikationen in den geförderten Wissenschaftsgebieten.

(2)[1962] Erwerb der Mitgliedschaft erfolgt durch Zuwahl durch die Mitglieder auf Vorschlag eines Mitglieds.

(3)[1962] Der Vorstand besteht aus drei Vorsitzenden und vier Beisitzern. Die drei Vorsitzenden bilden das *Präsidium*.

(4)[1962] Die Mitglieder wählen den Vorstand aus dem Kreise der Universitätsdozenten.

(5)[1962] Der Vorstand wählt das Präsidium aus seiner Mitte. Das Präsidium wählt den *geschäftsführenden Vorsitzenden* und den *Kassenwart* aus seiner Mitte.

(6)[1962] Es gibt keine Vorgaben über die Beschlußfähigkeit der Mitgliederversammlung.

(7)[1962] Satzungsänderungen werden von der Mitgliederversammlung mit einfacher Mehrheit beschlossen.

Auf der Mitgliederversammlung am 6. April 1967 in Oberwolfach wird die Satzung einstimmig geändert.[11] Die neue Satzung wird in Abschnitt 5 wiedergegeben.[12] Die neue Satzung enthält grundlegende Änderungen einiger Grundprinzipien des Vereins: der deutsche Wohnsitz als Voraussetzung für die Mitgliedschaft fällt weg, stattdessen wird der "wissenschaftliche Vortrag anläßlich einer Mitgliederversammlung" (§ 3) in die Satzung aufgenommen;[13] das Präsidium mit seinen drei Vorsitzenden wird abgeschafft und auf einen Vorsitzenden und einen Kassenwart reduziert; eine unscheinbare Änderung sollte bedeutende Nachwirkungen haben: konnte in der Satzung von 1962 die Mitgliederversammlung eine Satzungsänderung noch "mit einfacher Mehrheit" (1962, § 9) beschließen, so brauchte es nun die "einfache Mehrheit aller Mitglieder" (1967, § 11). Zusammenfassend:

---

[10]Vgl. DVMLG-Archiv A18.

[11]Vgl. DVMLG-Archiv A43.

[12]Es ist hervorzuheben, daß diese Satzung erstmals das Akronym DVMLG (statt DVLG) verwendet. Dieses Akronym wurde allerdings bereits früher regelmäßig für den Verein verwendet, sogar schon vor der förmlichen Gründung des Vereins; vgl. Brief von Arnold Schmidt an das Ministerium des Inneren v. 4. April 1961 (DVMLG-Archiv A10).

[13]Vgl. B. Löwe, Die frühe Mitgliederentwicklung der DVMLG, in diesem Bande.

(1)¹⁹⁶⁷ Voraussetzungen für die Mitgliedschaft sind eigene Publikationen in den geförderten Wissenschaftsgebieten.

(2)¹⁹⁶⁷ Erwerb der Mitgliedschaft erfolgt durch Zuwahl durch die Mitglieder auf schriftlichen Antrag eines Mitglieds nach einem wissenschaftlichen Vortrag.

(3)¹⁹⁶⁷ Der Vorstand besteht aus fünf Mitgliedern.

(4)¹⁹⁶⁷ Die Mitgliederversammlung wählt den Vorstand.

(5)¹⁹⁶⁷ Der Vorsitzende und der Kassenwart werden vom Vorstand aus seiner Mitte gewählt.[14]

(6)¹⁹⁶⁷ Die Mitgliederversammlung ist beschlußfähig, wenn mindestens ein Drittel der Mitglieder und zwei Vorstandsmitglieder anwesend sind.

(7)¹⁹⁶⁷ Satzungsänderungen werden von der Mitgliederversammlung mit einfacher Mehrheit aller Mitglieder beschlossen.

Die Vorgaben der Satzungen der Jahre 1962 und 1967 gingen davon aus, daß es sich bei der DVMLG um einen kleinen Verein handelte, dessen Mitglieder sich regelmäßig bei den jährlichen Vereinsveranstaltungen trafen. Insbesondere die Regelung der Zuwahl von Neumitgliedern, die nun einen wissenschaftlichen Vortrag vor der Mitgliederversammlung vor der Abstimmung über die Zuwahl vorsah, war vor dem Hintergrund dieses Vereinsideals zu sehen. Ende 1969 hatte die DVMLG 46 Mitglieder und die Zahl der Mitglieder wuchs schnell. In den 1970er Jahren wurde deutlich, daß die Strukturen des Vereins an die neue Situation angepaßt werden mußten. Auf der Mitgliederversammlung in Oberwolfach am 1. April 1971 wurde ein Meinungsbild über einige dieser Grundprinzipien erstellt: Neumitglieder sollten in Zukunft vom Vorstand aufgenommen werden, wenn kein Widerspruch der Mitglieder vorlag; der Vorstand sollte auf sechs Personen erweitert werden. Eine etwas kuriose Empfehlung, die nicht in der Satzung von 1972 aufgenommen wurde, ist diejenige für die Beschlußfähigkeit der Mitgliederversammlung:

> Die Beschlußfähigkeit liegt vor, wenn alle Mitglieder oder mehr als das Zehnfache des dekadischen Logarithmus aller Mitglieder anwesend sein.[15]

---

[14]Vgl. §7 zu einer Sonderregelung bei der Wahl des Vorsitzenden, wenn weniger als drei Vorstandsmitglieder anwesend sind.

[15]DVMLG-Archiv A54. Die Mitgliederversammlungen dieser Zeit in Oberwolfach dauerten oft mehr als vier Stunden und wurden erst nach Mitternacht beendet. Diese Sitzung endete laut Protokoll um 0:40 Uhr am 2. April 1971 nach fast fünf Stunden; der Tagesordnungspunkt *Satzungsänderung* war der letzte vor *Verschiedenes*.

Das Meinungsbild der Mitgliederversammlung 1971 wurde in konkrete Änderungsvorschläge umgesetzt, die auf der Mitgliederversammlung in Oberwolfach am 19. April 1972 mit 32 Stimmen bei einer Enthaltung angenommen wurden.[16] Allerdings hatte die DVMLG zu diesem Zeitpunkt bereits 70 Mitglieder, so daß diese 32 Stimmen nicht die Hälfte aller Mitglieder waren und sie somit nicht für eine Satzungsänderung ausreichten. Der Vorsitzende, Arnold Oberschelp, schrieb daher am 26. April 1972 die Mitglieder an und bat um zusätzliche Zustimmung von mindestens vier Mitgliedern, um den Beschluß satzungskonform zu machen.[17] Dem Sitzungsprotokoll (DVMLG-Archiv A56) liegt ein von Arnold Oberschelp und Klaus Potthoff unterzeichnetes Wahlprotokoll bei, in dem vermeldet wird, daß "23 Mitglieder geantwortet [haben] und zwar 22 mit Ja, ein Mitglied mit Enthaltung".

Einige der Beschlüsse des Jahres 1971 wurden nicht umgesetzt: z.B. wollte die Mitgliederversammlung im Jahre 1971 noch die Pflicht des wissenschaftlichen Vortrags aufrechterhalten; in der Satzung von 1972 wird diese nicht mehr erwähnt. Zudem enthält die neue Satzung eine Gemeinnützigkeitsklausel, da der Verein für eine im Jahre 1974 geplante Tagung in Kiel spendenfähig gemacht werden sollte. Zusammenfassend:

(1)$^{1972}$ Voraussetzungen für die Mitgliedschaft sind in der Regel eigene Publikationen in den geförderten Wissenschaftsgebieten.

(2)$^{1972}$ Erwerb der Mitgliedschaft erfolgt auf schriftlichen Antrag. Jedes Mitglied hat das Recht, eine Wahl auf der Mitgliederversammlung zu erzwingen. Findet dies nicht statt, so beschließt der Vorstand über die Aufnahme.

(3)$^{1972}$ Der Vorstand besteht aus sechs Mitgliedern.

(4)$^{1972}$ Die Mitgliederversammlung wählt den Vorstand.

(5)$^{1972}$ Der Vorsitzende wird von der Mitgliederversammlung aus dem Vorstand gewählt. Der Kassenwart wird vom Vorstand aus seiner Mitte gewählt.

(6)$^{1972}$ Die Mitgliederversammlung ist beschlußfähig, wenn mindestens ein Drittel der Mitglieder oder mindestens dreißig Mitglieder anwesend sind.

(7)$^{1972}$ Satzungsänderungen werden von der Mitgliederversammlung mit einfacher Mehrheit aller Mitglieder beschlossen. Die Möglichkeit der brieflichen Zustimmung wird explizit erwähnt.

Die Satzung von 1972 wird in Abschnitt 6 wiedergegeben.

---

[16] DVMLG-Archiv A56.

[17] Brief von Arnold Oberschelp an die Mitglieder der DVMLG v. 26. April 1972 (DVMLG-Archiv A57).

Mit dieser neuen Satzung beantragte Arnold Oberschelp die Gemeinnützigkeit beim Finanzamt Kiel, wurde aber informiert, daß die Formulierungen in der Satzung nicht ausreichten. Man passte die Satzung an, stimmte auf der Mitgliederversammlung in Oberwolfach am 12. April 1973 ab, wobei wiederum die "einfache Mehrheit aller Mitglieder" nicht erreicht wurde und bat die Mitglieder um schriftliche Abstimmung. Die Satzungsänderung wurde am 5. Juni 1973 angenommen.[18] Die Änderungen von § 12 und § 13 finden sich in Abschnitt 7.

In den folgenden Jahren fanden die Mitgliederversammlungen z.T. im Zusammenhang mit anderen Tagungen statt, so z.B. im Rahmen der Jahrestagung der Deutschen Mathematiker-Vereinigung (DMV) in München im Jahre 1976. Diese Veranstaltungen zogen weniger Mitglieder der DVMLG an als die jährlichen Oberwolfach-Tagungen und resultierten somit in nicht beschlußfähigen Mitgliederversammlungen. So waren die Mitgliederversammlungen in den Jahren 1973 und 1978 nicht beschlußfähig.[19] Auf der Mitgliederversammlung in Dortmund am 19. September 1980 und der Mitgliederversammlung in Bayreuth am 23. September 1983 wurde die ersatzlose Streichung des Quorums besprochen und in einen neuen Satzungsvorschlag eingearbeitet.[20] Inzwischen war allerdings auch die briefliche Abstimmung der Mitglieder über Satzungsänderungen nicht mehr einfach. Die erste briefliche Abstimmung erfolgte im Dezember 1982, aber

> [l]eider war jedoch auf dem Abstimmungszettel [die Frage zur Satzungsänderung] etwas unglücklich formuliert und ging eindeutig nur aus dem Protokoll der Mitgliederversammlung hervor. Es erscheint mir wegen der Wichtigkeit der Frage korrekt, diese Abstimmung noch einmal zu wiederholen.[21]

Die Abstimmung wurde im Februar 1983 wiederholt, erreichte aber nicht die "einfache Mehrheit aller Mitglieder".[22] Eine erneute Wiederholung im

---

[18]DVMLG-Archiv A62 enthält u.a. das Protokoll der Sitzung in Oberwolfach und das von Oberschelp und Potthoff unterzeichnete zusätzliche Wahlprotokoll. Vgl. auch den Brief von Arnold Oberschelp an den Vorstand der DVMLG v. 12. Juli 1973 (DVMLG-Archiv A61).
[19]Protokoll der Mitgliederversammlung der DVMLG am 12. April 1973 in Oberwolfach (Universitätsarchiv Freiburg B 0160/2) und das Protokoll der Mitgliederversammlung der DVMLG am 3. Oktober 1978 in Aachen (DVMLG-Archiv A84).
[20]DVMLG-Archiv A93 & A108.
[21]Brief von Michael M. Richter an die Mitglieder der DVMLG v. Dezember 1982 (DVMLG-Archiv A110).
[22]Es gab bei der Abstimmung drei Optionen (*Antrag I*, *Antrag II* und *Antrag III* und es fielen 3 Stimmen auf *Antrag I*, 49 Stimmen auf *Antrag II* und 19 Stimmen auf *Antrag III* bei einer Enthaltung (Brief von Michael M. Richter an die Mitglieder der DVMLG v. März 1983; DVMLG-Archiv A113). Das Amtsgericht Marburg antwortete: "Nach § 11 der Satzung können Satzungsänderungen mit einfacher Mehrheit aller Mitglieder beschlossen werden. Ob die erforderlichen Mehrheitsverhältnisse gegeben sind, kann ich von hier aus

April 1985 erreichte wiederum nicht die erforderliche Mehrheit.[23] Im Jahre 1986 setzte der neue Vorsitzende Klaus Potthoff den "mehrjährigen Versuch die Satzung zu ändern"[24] fort, überarbeitete die ersten Artikel der Satzung gründlich, um die Gemeinnützigkeit des Vereins zu erhalten und bat dann um schriftliche Abstimmung. Die Änderungen in § 1 bis 5 erhielten 70 Ja-Stimmen, eine Gegenstimme und zwei Enthaltungen; die Änderungen in § 9 (nach neuer Zählung § 12) erhielten 69 Ja-Stimmen, zwei Gegenstimmen und zwei Enthaltungen:

> Die DVMLG hat 132 Mitglieder. Daher ist die erforderliche Mehrheit für eine Satzungsänderung in beiden Fällen erreicht und die Satzungsänderungen sind beschlossen.[25]

Der Text der Satzung aus dem Jahre 1986 ist in Abschnitt 8 wiedergegeben.[26] Die meisten Vorgaben der Satzung von 1972 blieben erhalten; die wesentliche Änderung war die Beschlußfähigkeit der Mitgliederversammlung.

(6)$^{1986}$ Satzungsgemäß einberufene Mitgliederversammlungen sind grundsätzlich beschlußfähig.

Auf der Mitgliederversammlung in Dresden am 23. September 2000 wurden drei bedeutende Satzungsänderungen vorgeschlagen.[27]

Die weitreichendste Veränderung findet sich in § 6, dem Artikel, der die Aufnahme neuer Mitglieder beschreibt. In den ersten Jahren des Vereins hatte dieser Artikel zum Ziel, die Mitgliedschaft des Vereins selektiv zu halten und den Mitgliedern Kontrolle über die Zuwahl von Neumitgliedern zu gewähren: das Bestehen auf eigenen Publikationen, einem wissenschaftlichen Vortrag vor den Mitgliedern und das Veto-Recht jedes einzelnen Mitglieds, welches erzwingen konnte, dass über ein Aufnahmegesuch in der Mitgliederversammlung abgestimmt wird, sorgten dafür, daß die Mitgliedschaft

---

nicht beurteilen, da mir die Zahl der Gesamtmitglieder nicht bekannt ist." (Brief vom Amtsgericht Marburg an Michael M. Richter v. 18. November 1983; DVMLG-Archiv A114.)

[23] Brief vom Amtsgericht Marburg an Michael M. Richter v. 21. Mai 1985 (DVMLG-Archiv A122): "Nach Ihrem Rundbrief hat die Auszählung am 22.4.1985 ergeben, daß 54 Mitglieder sich für die Änderung—8 Mitglieder gegen die Änderung ausgesprochen haben. ... In Ihrem Schreiben ... teilten Sie mit, daß der Verein zum damaligen Zeitpunkt 132 Mitglieder zählte; eine Satzungsänderung hätte die Zustimmung ... von mindestens 67 Mitgliedern erfordert."

[24] Brief von Klaus Potthoff an die Mitglieder der DVMLG v. 29. August 1986 (DVMLG-Archiv A130).

[25] Protokoll über die Wahl zweier Vorstandsmitglieder und der Satzungsänderungen der DVMLG am 3. 12. 1986 (DVMLG-Archiv A136).

[26] DVMLG-Archiv NE 1996–2001 Verschiedenes.

[27] Vgl. Deutsche Vereinigung für Mathematische Logik und Grundlagenforschung der exakten Wissenschaften. Protokoll der Mitgliederversammlung. TU Dresden, 23. September 2000, 20 Uhr (DVMLG-Archiv NE 1996–2001 / laufende Korrespondenz).

schwer zu erlangen war. Die gesellschaftliche Stellung von Fachverbänden hatte sich am Anfang des neuen Jahrtausends verändert und man strebte eine offenere Mitgliedspolitik an. Die Vorschläge des Jahres 2000 schwächten die Forderung nach eigenen wissenschaftlichen Publikationen zu "wissenschaftlicher Arbeit" ab, was es dem Verein ermöglichte, Doktorandinnen und Doktoranden, die noch keine eigenen Veröffentlichungen hatten, in den Verein aufzunehmen. Das Recht einzelner Mitglieder, eine Abstimmung über Aufnahmegesuche zu erzwingen, verschwand; stattdessen wurde nun umgekehrt der Vorstand gezwungen, Ablehnungen von der Mitgliederversammlung bestätigen zu lassen. Zusätzlich wurde das Amt der oder des stellvertretenden Vorsitzenden in § 10 und 11 eingeführt.

Über die Satzungsänderung wurde auf der Mitgliederversammlung und durch briefliche Stimmabgabe abgestimmt. Für die Änderungen in § 6 gab es 71 Ja-Stimmen, fünf Gegenstimmen und eine Enthaltung; für die Änderungen in § 10 und 11 gab es 77 Ja-Stimmen und keine Gegenstimmen und Enthaltungen.[28] Der Änderungen des Textes sind in Abschnitt 9 wiedergegeben. Im Vergleich zu den vorigen Satzungen ändern sich drei der sieben Vorgaben:

(1)$^{2000}$ Voraussetzungen für die Mitgliedschaft ist wissenschaftliche Arbeit in den geförderten Wissenschaftsgebieten.

(2)$^{2000}$ Erwerb der Mitgliedschaft erfolgt auf schriftlichen Antrag durch Vorstandsbeschluß. Ablehnungen müssen von der Mitgliederversammlung bestätigt werden.

(5)$^{2000}$ Der Vorsitzende und der stellvertretende Vorsitzende werden von der Mitgliederversammlung aus dem Vorstand gewählt. Der Kassenwart wird vom Vorstand aus seiner Mitte gewählt.

## 3 Der Satzungsentwurf der Vorgängerinstitution (vermutlich 1954)

*Entwurf. Satzung der deutschen Vereinigung für mathematische Logik und für Grundlagenforschung der exakten Wissenschaften (DVLG).*

§ 1. Der Zweck der DVLG ist die Förderung der in ihrem Namen aufgeführten Wissenschaftszweige und im Zusammenhang hiermit insbesondere die Förderung des wissenschaftlichen Kontaktes der mit diesen Wissenschaftszweigen befaßten Forscher des In- und Auslandes.

§ 2. Ordentliches Mitglied kann jeder erwachsene Deutsche werden, gegen dessen Aufnahme in die Vereinigung nicht vom Vorstand Einspruch

---

[28]Vgl. Protokoll über die Stimmenauszählung zur Satzungsänderung der DVMLG v. 8. November 2000 (DVMLG-Archiv NE 1996–2001 / laufende Korrespondenz).

erhoben wird. Förderndes Mitglied kann jede juristische Person des In- und Auslandes werden, gegen deren Aufnahme nicht vom Vorstand Einspruch erhoben wird.

**§ 3.** Der Jahresbeitrag für ordentliche und für fördernde Mitglieder wird vom Vorstand den finanziellen Anforderungen gemäß bis spätestens zum Ende Oktober des Vorjahres festgesetzt. Ergeht hierzu kein besonderer Beschluß, so behält der Jahresbeitrag die Höhe, die er im Vorjahre hatte.

**§ 4.** Die Mitglieder des DVLG [sic!] können das "Archiv für mathematische Logik und Grundlagenforschung" (AMLG) zu einem ermäßigten Preise beziehen.

**§ 5.** Der Vorstand (Gesamtvorstand) besteht aus sechs Mitgliedern. Drei von ihnen bilden das Präsidium, die übrigen drei den "erweiterten Vorstand".

**§ 6.** Das Präsidium wird alle drei Jahre vom Gesamtvorstand aus den Mitgliedern des Gesamtvorstandes gewählt; jedoch bleibt das erste Präsidium für zwei Wahlperioden im Amt. Wiederwahl ist zulässig. Bei Ausscheiden eines Mitglieds des Präsidiums findet innerhalb von zwei Monaten eine Ersatzwahl durch den restlichen Gesamtvorstand, in der Regel aus seinen Mitgliedern, statt.

**§ 7.** Alle zwei Jahre werden zwei der drei Mitglieder des erweiterten Vorstandes neugewählt, und zwar bestimmt jeweils das Los diejenigen beiden Mitglieder, die ihre Mitgliedschaft zur Verfügung stellen. Wiederwahl ist zulässig. Die Mitglieder des erweiterten Vorstandes sollen in der Regel Dozenten einer Hochschule sein, die Mitglieder des Präsidiums m ü s s e n diese Eigenschaft besitzen.

**§ 8.** Die Ergänzungswahl zum erweiternden Vorstand erfolgt in der Regel auf einer Mitgliederversammlung; eine solche ist hierzu beschlußfähig, sofern mindestens sechs Mitglieder anwesend sind. Falls eine solche Versammlung bis zum Ende des Jahres, in dem die in § 7 genannte Zweijahresfrist abläuft, nicht zustandekommt, so ist schriftliche Wahl aufgrund vorher schriftlich beim geschäftsführenden Vorsitzenden eingereichter Vorschläge zulässig. Das Präsidium macht in diesem Falle den Mitgliedern Mitteilung von der bevorstehenden Wahl. Daraufhin können innerhalb von zwei Wochen Wahlvorschläge eingereicht werden; jeder Kandidat muß von mindestens drei Mitgliedern vorgeschlagen sein. Die schriftliche Wahl erfolgt durch Ausfüllung der vom Präsidium den Mitgliedern übersandten Stimmzetteln innerhalb einer dabei anzugebenden angemessenen Frist.

**§ 8.** [sic!] Das Präsidium bestimmt zwei seiner Mitglieder als Vorsitzende, davon eines als geschäftsführenden Vorsitzenden, und ein. weiteres Mitglied als Beitragswart (Sekretär der Vereinigung?). Das Präsidium führt die Geschäfte der Vereinigung; der geschäftsführende Vorsitzende besitzt die

Unterschriftsbefugnis. Eine Angelegenheit, in der innerhalb des Präsidiums trotz eingehender (mündlicher oder schriftlicher) Beratung keine Einstimmigkeit erzielt wird, wird vom geschäftsführenden Vorsitzenden dem Gesamtvorstand zur Entscheidung vorgelegt.

§ 9. Die Mitgliederversammlungen werden vom geschäftsführenden Vorsitzenden in der Regel alle zwei Jahre einberufen. Auf der Mitgliederversammlung legt das Präsidium einen Tätigkeitsbericht vor und legt über die Verwendung der Beitragsgelder der Vereinigung Rechenschaft ab.

§ 10. Eine Änderung der Satzung bedarf einer Zweidrittelmehrheit in einer Mitgliederversammlung, auf der mindestens sechs Mitglieder und mindestens ein Drittel der Mitglieder anwesend sind. Dasselbe gilt für die Auflösung der Vereinigung.

## 4 Die Gründungssatzung (1962)

*Satzung der deutschen Vereinigung für mathematische Logik und für Grundlagenforschung der exakten Wissenschaften (DVLG).*

§ 1. Die DVLG hat sich die Förderung der in ihrem Namen aufgeführten Wissenschaftszweige zum Ziele gesetzt; insbesondere bezweckt sie die Pflege des wissenschaftlichen Kontaktes der mit diesen Wissenschaftszweigen befaßten Forscher und Institutionen des In- und Auslandes.

§ 2. Die DVLG wird beim Amtsgericht in Marburg/Lahn eingetragen; ihr Sitz ist Marburg/Lahn.

§ 3. Ordentliche Mitglieder können nach Maßgabe des nachfolgenden Absatzes alle Personen werden, die in der Bundesrepublik Deutschland oder in Westberlin ihren Wohnsitz haben und auf den Gebieten, deren Förderung die Vereinigung bezweckt, mit eigenen Publikationen hervorgetreten sind.

Der Erwerb der Mitgliedschaft erfolgt durch Zuwahl auf Vorschlag eines ordentlichen Mitgliedes. Die Zuwahl gilt als nicht erfolgt, wenn sich weniger als zwei Drittel der auf der Mitgliederversammlung anwesenden Mitglieder für sie ausgesprochen haben, Die Wahl ist geheim vorzunehmen, wenn ein Mitglied dies verlangt.

Förderndes Mitglied kann jede juristische Person des In- oder Auslandes werden, solange gegen diese Mitgliedschaft nicht vom Vorstand Einspruch erhoben wird.

§ 4. Ein Mitglied kann jederzeit seinen Austritt erklären, ist jedoch verpflichtet, den Mitgliedsbeitrag für das laufende Jahr zu zahlen. Die Mitgliedschaft endet auch durch den Tod des Mitgliedes.[29]

§ 5. Verstößt ein ordentliches Mitglied in grober Weise gegen den Zweck oder das Ansehen der DVLG, so kann es auf Antrag des Vorstandes mit einfacher Mehrheit ausgeschlossen werden.

---
[29] Der letzte Satz fehlt in A1.

§ 6. Der Vorstand besteht aus drei Vorsitzenden und vier Beisitzern. Die drei Vorsitzenden bilden das Präsidium. Sie wählen unter sich den geschäftsführenden Vorsitzenden und den Kassenwart.

Das Präsidium führt die Geschäfte der Vereinigung; der geschäftsführende Vorsitzende besitzt Unterschriftsbefugnis im Namen des Präsidiums.

Die Verfügung über die Kasse (das Konto) der Vereinigung steht dem Präsidium zu. Unterschriftsberechtigt ist der Kassenwart.

Das Präsidium faßt seine Beschlüsse einstimmig.

Der geschäftsführende Vorsitzende kann die Zustimmung der anderen Vorsitzenden schriftlich einholen. Kommt in einer Frage keine Einstimmigkeit zustande, so kann der geschäftsführende Vorsitzende eine (mündliche oder schriftliche) Entscheidung des Vorstandes herbeiführen. Er muß dies, wenn ein Vorsitzender die Entscheidung des Vorstandes verlangt.

§ 7. Der Vorstand faßt seine Beschlüsse mit einfacher Mehrheit. Der Vorstand wählt das Präsidium aus seiner Mitte für drei Jahre, Wiederwahl ist zulässig. Das erste Präsidium wird für zwei Wahlperioden gewählt. Scheidet ein Mitglied des Präsidiums aus, so hat der restliche Vorstand innerhalb von zwei Monaten einen weiteren Vorsitzenden als Ersatz zu wählen.

Alle drei Jahre scheiden nach der Wahl des Präsidiums die beiden am längsten dem Vorstand angehörenden Beisitzer aus dem Vorstande aus, Wiederwahl ist zulässig. Im Zweifelsfalle entscheidet das Los.[30]

§ 8. Die Mitgliederversammlung besetzt durch Wahl die freigewordenen Vorstandsstellen aus dem Kreise derjenigen Mitglieder, die Universitätsdozenten sind, neu. Ein solches Mitglied gilt als gewählt, wenn es die absolute Mehrheit der Anwesenden auf sich vereinigt. Kommt keine absolute Mehrheit zustande, so findet eine Stichwahl zwischen denjenigen beiden Mitgliedern statt, die die relativ meisten Stimmen auf sich vereinigt haben. Als gewählt gilt bei der Stichwahl, wer die meisten Stimmen erhält.[31] Die Mitgliederversammlung findet mindestens alle 3 Jahre statt.[32]

Der erste Vorstand wird en bloc mit einfacher Mehrheit gewählt.

§ 9. Satzungsänderungen werden auf Vorschlag des Präsidiums oder auf Mehrheitsvorschlag des Vorstandes von der Mitgliederversammlung mit einfacher Mehrheit beschlossen.

§ 10. Der Mitgliedsbeitrag wird mit dem Beginn des Kalenderjahres fällig. Eine Änderung seiner Höhe bestimmt die Mitgliederversammlung auf Vorschlag des Präsidiums. Bleibt ein Mitglied mit der Zahlung auch nach einmaliger Mahnung im Rückstand, so kann es nach Ablauf von drei Monaten nach Abgang der Mahnung auf Beschluß des Präsidiums ausgeschlossen werden.

---

[30]Der letzte Satz fehlt in A1.
[31]A1 hat "Gewählt" statt "Als gewählt".
[32]Der letzte Satz fehlt in A1.

§ 11. Die Mitgliederversammlungen werden vom geschäftsführenden Vorsitzenden einberufen. Über die Beschlüsse der Mitgliederversammlung ist ein Protokoll anzufertigen. Es ist von dem geschäftsführenden Vorsitzenden zu unterschreiben.[33]

§ 12. Die Auflösung der Vereinigung kann auf Vorschlag des Vorstandes mit Dreiviertel-Mehrheit von der Mitgliederversammlung beschlossen werden. Im Falle der Auflösung fällt das Vereinsvermögen der Deutschen Forschungsgemeinschaft zu.

§ 13. Die letzte Bestimmung des § 12 kann von der Mitgliederversammlung nicht geändert werden.

§ 14. Das Geschäftsjahr des Vereins ist das Kalenderjahr.[34]

## 5 Die Satzung von 1967

*Satzung der Deutschen Vereinigung für mathematische Logik und für Grundlagenforschung der exakten Wissenschaften (DVMLG). Geänderte Fassung, beschlossen am 6. April 1967.*

§ 1 & 2. *Unverändert.*

§ 3. Ordentliches Mitglied kann jede Person nach Maßgabe des nachfolgenden Absatzes werden, in der Regel sofern sie auf den Gebieten, deren Förderung die Vereinigung bezweckt, mit eigenen Publikationen hervorgetreten ist. Der Erwerb der Mitgliedschaft erfolgt auf schriftlichen Antrag eines ordentlichen Mitgliedes gemäß dem Verfahren in § 9 durch Zuwahl in einer Mitgliederversammlung. Die Zuwahl eines Mitgliedes gilt als erfolgt, wenn auf der Mitgliederversammlung die Anzahl der Ja-Stimmen größer als die Hälfte der Anwesenden und mindestens doppelt so groß wie die Anzahl der Nein-Stimmen ist. Ein für die Aufnahme vorgeschlagener Kandidat soll anläßlich einer Mitgliederversammlung vor der Abstimmung über die Aufnahme einen wissenschaftlichen Vortrag halten, Ausnahmen hiervon müssen im Antrag begründet werden und bedürfen der Zustimmung von zwei Dritteln der Anwesenden Mitglieder.

Förderndes Mitglied kann jede juristische Person des In- oder Auslandes werden, solange gegen diese Mitgliedschaft nicht vom Vorstand Einspruch erhoben wird.

§ 4. *Unverändert (s. Abschnitt 4).*

§ 5. *Die Worte "einfache Mehrheit" wurden ersetzt durch "Zweidrittelmehrheit" (s. Abschnitt 4).*

§ 6. Der Vorstand besteht aus 5 ordentlichen Mitgliedern. Jedes Jahr scheidet das Vorstandsmitglied, dessen Wahl am längsten zurückliegt, anläßlich der Mitgliederversammlung aus. Die Mitgliederversammlung ergänzt

---
[33] Die letzten zwei Sätze fehlen in A1.
[34] § 14 fehlt in A1.

den Vorstand durch Zuwahl eines Mitgliedes. Die Zuwahl eines Vorstandsmitgliedes gilt als erfolgt, wenn es die absolute Mehrheit der Anwesenden auf sich vereinigt. Kommt keine absolute Mehrheit zustande, so findet eine Stichwahl zwischen denjenigen beiden Mitgliedern statt, die die relativ meisten Stimmen auf sich vereinigt haben. Als gewählt gilt bei der Stichwahl, wer die meisten Stimmen erhält. In Zweifelsfällen entscheidet das Los. Wiederwahl ist unbeschränkt zulässig. Wenn in einem Jahr keine Mitgliederversammlung stattfindet, bleibt das turnusmäßig zu ersetzende Vorstandsmitglied bis zur nächsten Mitgliederversammlung im Amt. Die Wahl seines Nachfolgers, dessen Amtszeit sich entsprechend verkürzt, ist gleichzeitig mit der nächsten Ergänzung des Vorstandes nachzuholen.

Scheidet ein Vorstandsmitglied vorzeitig aus, so ist auf der nächsten Mitgliederversammlung für den Rest seiner Amtszeit ein neues Vorstandsmitglied zu wählen.

§ 7. Ein Mitglied des Vorstandes ist Vorsitzender der Vereinigung. Die Amtszeit des Vorsitzenden beträgt rund zwei Jahre. Wiederwahl ist unbeschränkt zulässig. Wenn der Vorsitzende vor Ablauf dieser Periode als Vorstandsmitglied ausscheidet, ohne wieder in den Vorstand gewählt zu werden, so scheidet er auch als Vorsitzender aus.

Wenn in dem Jahr, in dem seine Amtsperiode abläuft, keine Mitgliederversammlung stattfindet, bleibt er bis zur nächsten Mitgliederversammlung im Amt. Der Vorsitzende wird wie folgt gewählt. Falls nach der Ergänzung des Vorstandes mindestens drei Vorstandsmitglieder anwesend sind, wählt der Vorstand in unmittelbarem Anschluß an die Mitgliederversammlung den Vorsitzenden aus seiner Mitte. Ein Vorstandsmitglied gilt als zum Vorsitzenden gewählt, wenn es mindestens drei Stimmen auf sich vereinigt. Kommt in drei Wahlgängen keine Entscheidung zustande, so wird in seiner zu diesem Zweck innerhalb 24 Stunden zu eröffnenden Mitgliederversammlung der Vorsitzende aus der Mitte des Vorstandes gewählt. Das Wahlverfahren ist dasselbe wie bei der Wahl eines Vorstandsmitgliedes. Falls dagegen nach de Ergänzung des Vorstandes weniger als drei Vorstandsmitglieder anwesend sind, wählt die Mitgliederversammlung, in der die Ergänzung des Vorstandes vorgenommen wurde, den Vorsitzenden nach dem obigen Verfahren.

§ 8. Der Vorstand führt die Geschäfte der Vereinigung. Der Vorsitze besitzt Unterschriftsbefugnis im Namen des Vorstandes. Er vertritt die Vereinigung gerichtlich und außergerichtlich.

Der Vorstand faßt seine Beschlüsse mit Mehrheit. Der Vorsitzende kann die Entscheidung der Vorstandsmitglieder schriftlich einholen. Der Vorstand regelt unter sich die Frage der Stellvertretung des Vorsitzenden.

Der Vorstand wählt mit einfacher Mehrheit einen Kassenwart aus dem Kreis der Vorstandsmitglieder. Die Verfügung über die Kasse (das Konto) der Vereinigung steht dem Vorstand zu. Unterschriftsberechtigt hierzu ist der Kassenwart.

Der Vorstand verteilt einzelne Aufgaben unter sich. Er kann Mitglieder der Vereinigung mit bestimmten Aufgaben befristet betrauen.

§ 9. Die Mitgliederversammlung werden vom Vorsitzenden in der Regel jährlich, jedoch mindestens nach zwei Jahren einberufen. Sie sollen nach Möglichkeit anläßlich wissenschaftlicher Tagungen stattfinden.

Auf jeder Mitgliederversammlung sollte der Termin für die nächste Mitgliederversammlung festgelegt werden. Eine Mitgliederversammlung muß spätestens ein Vierteljahr vorher angekündigt werden. Bis spätestens 8 Wochen vor einer Mitgliederversammlung kann jedes Mitglied Anträge für die Tagesordnung dem vorsitzenden einreichen. Anträge zur Aufnahme neuer Mitglieder sind schriftlich zu begründen.

Spätestens 6 Wochen vor einer Mitgliederversammlung erhalten die Mitglieder einen Entwurf der Tagesordnung, der alle Anträge der Mitglieder berücksichtigt. Zugleich sind den Mitgliedern die Begründungen zu den Anträgen auf Aufnahme neuer Mitglieder mitzuteilen.

In einer Mitgliederversammlung kann eine Erweiterung der Tagungsordnung mit Zweidrittelmehrheit beschlossen werden. Über die Beschlüsse der Mitgliederversammlung wird ein Protokoll angefertigt. Dieses wird am Ende der Mitgliederversammlung verlesen und genehmigt. Es wird vom Protokollführer und vom Vorsitzenden, der die Mitgliederversammlung geleitet hat, unterzeichnet. Eine Abschrift des Protokolls wird jedem Mitglied innerhalb von drei Monaten zugesandt. Gegebenenfalls ist bei dieser Gelegenheit der Name des neuen Vorsitzenden zu nennen.

Eine Mitgliederversammlung ist beschlußfähig, wenn mindestens ein Drittel der Mitglieder und zwei Vorstandsmitglieder anwesend sind.

Alle Personalwahlen sind geheim vorzunehmen.

§ 10. *Die Worte "auf Vorschlag des Präsidiums" wurden ersetzt durch "auf Vorschlag des Vorstandes" (s. Abschnitt 4).*

§ 11. Satzungsänderungen können auf Vorschlag des Vorstandes oder Vorschlag der Mehrheit der anwesenden Mitglieder einer Mitgliederversammlung mit einfacher Mehrheit aller Mitglieder beschlossen werden,

**§ 12 bis 14.** *Unverändert (s. Abschnitt 4).*

§ 15. (Übergangsbestimmungen) Die Satzung tritt mit der Eintragung beim Amtsgericht Marburg/Lahn in Kraft. Der amtierende Vorstand bleibt bis zur nächsten Mitgliederversammlung, die im Frühjahr 1968 erfolgen muß, im Amt. Auf dieser Mitgliederversammlung wird ein neuer Vorstand gemäß dem Verfahren von § 6 gewählt.

Die Mitgliederversammlung wählt aus dem Kreise des bisherigen Vorstandes nacheinander vier Mitglieder in den neuen Vorstand und anschließend aus dem Kreise der gesamten Mitgliederschaft ein weiteres Vorstandsmitglied.

Das Los entscheidet darüber, bei welchen der vier zuerst gewählten Vorstandsmitglieder die Wahl als ein, zwei, drei oder vier Jahre zurückliegend im Sinne von §6 gelten soll.

## 6 Die Satzung von 1972

*Satzung der Deutschen Vereinigung für mathematische Logik und für Grundlagenforschung der exakten Wissenschaften (DVMLG).*

§ 1. Die DVMLG hat sich die Förderung der in ihrem Namen aufgeführten Wissenschaftszweige zum Ziel gesetzt, insbesondere bezweckt sie die Pflege des wissenschaftlichen Kontaktes der mit diesen Wissenschaftszweigen befaßten Forscher und Institutionen des In- und Auslandes. Sie verfolgt ausschließlich und unmittelbar gemeinnützige Zwecke (im Sinne der Gemeinnützigkeitsordnung vom 24.12.1953). Etwaige Gewinne dürfen nur für die satzungsgemäßen Zwecke verwendet werden. Die Mitglieder erhalten keine Gewinnanteile und in ihrer Eigenschaft als Mitglieder auch sonst keine Zuwendung aus Mitteln des Vereins. Es darf keine Person durch Verwaltungsausgaben, die den Zwecken des Vereins fremd sind, oder durch unverhältnismäßig hohe Vergütungen begünstigt werden.

§ 2. Die DVMLG wird beim Amtsgericht in Marburg/Lahn eingetragen; ihr Sitz ist Marburg/Lahn.

§ 3. Ordentliches Mitglied kann jede Person werden, sofern sie in der Regel auf den Gebieten, deren Förderung die Vereinigung bezweckt, mit eigenen Publikationen hervorgetreten ist. Der Erwerb der Mitgliedschaft erfolgt auf schriftlichen Antrag eines Mitgliedes oder des Bewerbers. Über die Aufnahme bzw. Nichtaufnahme beschließt der Vorstand nach schriftlicher Unterrichtung der Mitglieder. Auf Verlangen eines Mitgliedes entscheidet die Mitgliederversammlung. Förderndes Mitglied kann jede juristische Person des In- und Auslandes werden, solange gegen diese Mitgliedschaft nicht vom Vorstand Einspruch erhoben wird.

§ 4. Ein Mitglied kann jederzeit seinen Austritt erklären, ist jedoch verpflichtet, den Mitgliedsbeitrag für das laufende Jahr zu zahlen. Die Mitgliedschaft endet durch den Tod des Mitgliedes.

§ 5. Verstößt ein ordentliches Mitglied in grober Weise gegen den Zweck oder das Ansehen der DVMLG, so kann es auf den Antrag des Vorstandes mit Zweidrittelmehrheit der Mitgliederversammlung ausgeschlossen werden.

§ 6. Der Vorstand besteht aus 6 ordentlichen Mitgliedern. Alle zwei Jahre scheiden die beiden Vorstandsmitglieder, deren Wahl am längsten zurückliegt, anläßlich einer Mitgliederversammlung aus. Die Mitgliederversammlung ergänzt den Vorstand durch Zuwahl zweier Mitglieder.

Scheidet ein Vorstandsmitglied vorzeitig aus, so ist auf der nächsten Mitgliederversammlung für den Rest seiner Amtszeit ein neues Vorstandsmitglied zu wählen.

§ 7. Ein Mitglied des Vorstandes ist Vorsitzender der Vereinigung. Die Amtszeit des Vorsitzenden beträgt zwei Jahre. Er wird von der Mitgliederversammlung gewählt.

§ 8. Der Vorstand führt die Geschäfte der Vereinigung. Der Vorsitzende besitzt Unterschriftsbefugnis im Namen des Vorstandes. Er vertritt die Vereinigung gerichtlich und außergerichtlich. Der Vorstand faßt seine Beschlüsse mit Mehrheit. Bei Stimmengleichheit entscheidet die Stimme des Vorsitzenden. Der Vorsitzende kann die Entscheidung der Vorstandsmitglieder schriftlich einholen. Der Vorstand regelt unter sich die Frage der Stellvertretung des Vorsitzenden.

Der Vorstand wähl mit einfacher Mehrheit einen Kassenwart aus dem Kreis der Vorstandsmitglieder. Die Verfügung über die Kasse (das Konto) der Vereinigung steht dem Vorstand zu. Unterschriftsberechtigt hierzu ist der Kassenwart. Der Vorstand verteilt einzelne Aufgaben unter sich. Er kann Mitglieder der Vereinigung mit bestimmten Aufgaben befristet betrauen.

§ 9. Die Mitgliederversammlungen werden vom Vorsitzenden mindestens alle zwei Jahre einberufen. Sie sollen nach Möglichkeit, anläßlich wissenschaftlicher Tagungen stattfinden.

Eine Mitgliederversammlung muß spätestens ein Vierteljahr vorher angekündigt werden. Bis spätestens vier Wochen vor einer Mitgliederversammlung kann Jedes Mitglied Anträge für die Tagesordnung dem Vorsitzenden einreichen. Spätestens zwei Wochen vor einer Mitgliederversammlung erhalten die Mitglieder einen Entwurf der Tagesordnung, der alle Anträge der Mitglieder berücksichtigt. Der Vorsitzende eröffnet die Mitgliederversammlung und läßt einen Versammlungsleiter wählen. Der Mitgliederversammlung obliegt die Wahl eines Kassenprüfers.

In einer Mitgliederversammlung kann eine Erweiterung der Tagungsordnung mit Zweidrittelmehrheit beschlossen werden. Anträge auf Satzungsänderungen und auf Abwahl von Vorstandsmitgliedern müssen jedoch in dem Entwurf der Tagesordnung angekündigt werden.

Über die Beschlüsse der Mitgliederversammlung wird ein Protokoll angefertigt. Dies wird am Ende der Mitgliederversammlung verlesen und genehmigt. Es wird von Protokollführer und vom Versammlungsleiter unterzeichnet. Eine Abschrift des Protokolls wird jedem Mitglied innerhalb von drei Monaten zugesandt.

Eine Mitgliederversammlung ist beschlußfähig, wenn mindestens ein Drittel der Mitglieder oder mindestens dreißig Mitglieder anwesend sind.

Ist eine Mitgliederversammlung nicht beschlußfähig, so kann der Vorstand Anträge zu den in der Tagesordnung genannten Punkten brieflich zur Abstimmung stellen.

Die Anzahl der abgegebenen Stimmen muß der Mindestanzahl entsprechen, die für die Beschlußfähigkeit festgesetzt ist. Alle Personalwahlen sind geheim vorzunehmen.

§ 10. Der Mitgliederbeitrag wird mit dem Beginn des Kalenderjahres fällig. Eine Änderung seiner Höhe bestimmt die Mitgliederversammlung auf Vorschlag des Vorstandes. Bleibt ein Mitglied mit der Zahlung auch nach zweimaliger Mahnung im Rückstand, so kann es nach Ablauf von drei Monaten nach Erhalt der zweiten Mahnung auf Beschluß des Vorstandes ausgeschlossen werden.

§ 11. Satzungsänderungen können auf Vorschlag des Vorstandes oder Vorschlag der Mehrheit der anwesenden Mitglieder einer Mitgliederversammlung mit einfacher Mehrheit aller Mitglieder beschlossen werden. Falls die in der Mitgliederversammlung gegebenen Stimmen zu einer Entscheidung nicht ausreichen, so sind die Stimmen der abwesenden Mitglieder innerhalb einer Woche mit vierwöchiger Frist brieflich einzuholen.

§ 12. Die Auflösung der Vereinigung kann auf Vorschlag des Vorstandes mit Dreiviertelmehrheit von der Mitgliederversammlung beschlossen werden. im Falle der Auflösung fällt das Vereinsvermögen der Deutschen Forschungsgemeinschaft zu.

§ 13. Die letzte Bestimmung des § 12 kann von der Mitgliederversammlung nicht geändert werden.

§ 14. Geschäftsjahr des Vereins ist das Kalenderjahr.

§ 15. (Übergangsbestimmungen) Die Satzung tritt mit der Eintragung beim Amtsgericht Marburg/Lahn in Kraft. Bereits auf der Mitgliederversammlung im April 1972 werden zwei neue Vorstandsmitglieder gemäß dieser neuen Satzung gewählt.

## 7 Die Satzung von 1973

*Satzung der Deutschen Vereinigung für mathematische Logik und für Grundlagenforschung der exakten Wissenschaften (DVMLG).*

§ 1 bis 11. *Unverändert (s. Abschnitt 6).*

§ 12. Die Auflösung der Vereinigung kann auf Vorschlag des Vorstandes mit Dreiviertelmehrheit von der Mitgliederversammlung beschlossen werden. im Falle der Auflösung fällt das Vereinsvermögen der Deutschen Forschungsgemeinschaft zu.

Gleiches gilt im Falle der Aufhebung des Vereins und für den Wegfall des bisherigen Zwecks, ohne daß ein anderer steuerbegünstigter Zweck bestimmt wird. In jedem der drei oben genannten Fälle ist das Vereinsvermögen ausschließlich und unmittelbar für die Zwecke der Förderung der Wissenschaften zu verwenden.

§ 13. Die Mitgliederversammlung kann keine Änderung der Satzung vornehmen, die der Zweckbestimmung des Vereins und des Vereinsvermögens, die Wissenschaft zu fördern, zuwiderläuft.

§ 14 bis 15. *Unverändert (s. Abschnitt 6).*

## 8 Die Satzung von 1986

*Deutsche Vereinigung für Mathematische Logik und Grundlagenforschung der exakten Wissenschaften (DVMLG). Satzung Fassung vom Dezember 1986.*[35]

§ 1. Die Deutsche Vereinigung für mathematische Logik und Grundlagenforschung der exakten Wissenschaften (DVMLG) (e.V.) mit Sitz in Marburg/Lahn verfolgt ausschließlich und unmittelbar gemeinnützige Zwecke im Sinne des Abschnitts „Steuerbegünstigte Zwecke" der Abgabenordnung. Zweck des Vereins ist die Förderung der in ihrem Namen aufgeführten Wissenschaftszweige. Der Satzungszweck wird verwirklicht insbesondere durch die Pflege des wissenschaftlichen Kontaktes der mit diesen Wissenschaftszweigen befaßten Forscher und Institutionen des In- und Auslandes, durch die Organisation von wissenschaftlichen Veranstaltungen und durch die Herausgabe wissenschaftlicher Informationsschriften.

§ 2. Der Verein ist selbstlos tätig; er verfolgt nicht in erster Linie eigenwirtschaftliche Zwecke

§ 3. Mittel des Vereins dürfen nur für die satzungsgemäßen Zwecke verwendet werden. Die Mitglieder erhalten keine Zuwendung aus Mittel des Vereins.

§ 4. Es darf keine Person durch Ausgaben, die dem Zweck der Körperschaft fremd sind, oder durch unverhältnismäßig hohe Vergütungen begünstigt werden.

§ 5. Bei Auflösung oder Aufhebung des Vereins oder bei Wegfall seines bisherigen Zwecks fällt das Vermögen des Vereins an die Deutsche Forschungsgemeinschaft, die es unmittelbar und ausschließlich für gemeinnützige, mildtätige oder kirchliche Zwecke zu verwenden hat.

§ 6. Ordentliches Mitglied kann jede Person werden, sofern sie in der Regel auf den Gebieten, deren Förderung die Vereinigung bezweckt, mit eigenen Publikationen hervorgetreten ist. Der Erwerb der Mitgliedschaft erfolgt auf schriftlichen Antrag eines Mitgliedes oder des Bewerbers. Über die Aufnahme bzw. Nichtaufnahme beschließt der Vorstand nach schriftlicher

---

[35]Das zweite "für" im Namen der Gesellschaft fehlt in der Satzung von 1986 und somit in der derzeit gültigen Satzung. Der im Vereinsregister eingetragene Name der Vereinigung ist weiterhin "Deutsche Vereinigung für mathematische Logik und *für* Grundlagenforschung der exakten Wissenschaften" (vgl. Brief vom Amtsgericht Marburg an Benedikt Löwe v. 18. Januar 2013).

Unterrichtung der Mitglieder. Auf Verlangen eines Mitgliedes entscheidet die Mitgliederversammlung. Förderndes Mitglied kann jede juristische Person des In- und Auslandes werden, solange gegen diese Mitgliedschaft nicht vom Vorstand Einspruch erhoben wird.

§ 7. Ein Mitglied kann jederzeit seinen Austritt erklären, ist jedoch verpflichtet, den Mitgliedsbeitrag für das laufende Jahr zu zahlen. Die Mitgliedschaft endet durch den Tod des Mitgliedes.

§ 8. Verstößt ein ordentliches Mitglied in grober Weise gegen den Zweck oder das Ansehen der DVMLG, so kann es auf den Antrag des Vorstandes mit Zweidrittelmehrheit der Mitgliederversammlung ausgeschlossen werden.

§ 9. Der Vorstand besteht aus 6 ordentlichen Mitgliedern. Alle zwei Jahre scheiden die beiden Vorstandsmitglieder, deren Wahl am längsten zurückliegt, anläßlich einer Mitgliederversammlung aus. Die Mitgliederversammlung ergänzt den Vorstand durch Zuwahl zweier Mitglieder.

Scheidet ein Vorstandsmitglied vorzeitig aus, so ist auf der nächsten Mitgliederversammlung für den Rest seiner Amtszeit ein neues Vorstandsmitglied zu wählen.

§ 10. Ein Mitglied des Vorstandes ist Vorsitzender der Vereinigung. Die Amtszeit des Vorsitzenden beträgt zwei Jahre. Er wird von der Mitgliederversammlung gewählt.

§ 11. Der Vorstand führt die Geschäfte der Vereinigung. Der Vorsitzende besitzt Unterschriftsbefugnis im Namen des Vorstandes. Er vertritt die Vereinigung gerichtlich und außergerichtlich. Der Vorstand faßt seine Beschlüsse mit Mehrheit. Bei Stimmengleichheit entscheidet die Stimme des Vorsitzenden. Der Vorsitzende kann die Entscheidung der Vorstandsmitglieder schriftlich einholen. Der Vorstand regelt unter sich die Frage der Stellvertretung des Vorsitzenden.

Der Vorstand wählt mit einfacher Mehrheit einen Kassenwart aus dem Kreis der Vorstandsmitglieder. Die Verfügung über die Kasse (das Konto) der Vereinigung steht dem Vorstand zu. Unterschriftsberechtigt hierzu ist der Kassenwart. Der Vorstand verteilt einzelne Aufgaben unter sich. Er kann Mitglieder der Vereinigung mit bestimmten Aufgaben befristet betrauen.

§ 12. Die Mitgliederversammlungen werden vom Vorsitzenden mindestens alle zwei Jahre einberufen. Sie sollen nach Möglichkeit, anläßlich wissenschaftlicher Tagungen stattfinden. Eine Mitgliederversammlung muß spätestens ein Vierteljahr vorher angekündigt werden. Bis spätestens vier Wochen vor einer Mitgliederversammlung kann jedes Mitglied Anträge für die Tagesordnung dem Vorsitzenden einreichen. Spätestens zwei Wochen vor einer Mitgliederversammlung erhalten die Mitglieder einen Entwurf der Tagesordnung, der alle Anträge der Mitglieder berücksichtigt. Der Vorsitzende eröffnet die Mitgliederversammlung und läßt einen Versammlungsleiter wählen. Der Mitgliederversammlung obliegt die Wahl eines Kassenprüfers.

In einer Mitgliederversammlung kann eine Erweiterung der Tagungsordnung mit Zweidrittelmehrheit beschlossen werden. Anträge auf Satzungsänderungen, Auflösung des Vereins und auf Abwahl von Vorstandsmitgliedern müssen jedoch in dem Entwurf der Tagesordnung angekündigt werden.

Über die Beschlüsse der Mitgliederversammlung wird ein Protokoll angefertigt. Dies wird am Ende der Mitgliederversammlung verlesen und genehmigt. Es wird von Protokollführer und vom Versammlungsleiter unterzeichnet. Eine Abschrift des Protokolls wird jedem Mitglied innerhalb von drei Monaten zugesandt.

Die Mitgliederversammlung kann mit einfacher Mehrheit beschließen, einzelne Punkte brieflich zur Abstimmung zu stellen.

§ 13. Der Mitgliederbeitrag wird mit dem Beginn des Kalenderjahres fällig. Eine Änderung seiner Höhe bestimmt die Mitgliederversammlung auf Vorschlag des Vorstandes. Bleibt ein Mitglied mit der Zahlung auch nach zweimonatiger Mahnung im Rückstand, so kann es nach Ablauf von drei Monaten nach Erhalt der zweiten Mahnung auf Beschluß des Vorstandes ausgeschlossen werden.

§ 14. Satzungsänderungen können auf Vorschlag des Vorstandes oder Vorschlag der Mehrheit der anwesenden Mitglieder einer Mitgliederversammlung mit einfacher Mehrheit aller Mitglieder beschlossen werden.

Falls die in der Mitgliederversammlung gegebenen Stimmen zu einer Entscheidung nicht ausreichen, so sind die Stimmen der abwesenden Mitglieder innerhalb einer Woche mit vierwöchiger Frist brieflich einzuholen.

§ 15. Die Auflösung der Vereinigung kann auf Vorschlag des Vorstandes mit Dreivierteimehrheit von der Mitgliederversammlung beschlossen werden. im Falle der Auflösung fällt das Vereinsvermögen der Deutschen Forschungsgemeinschaft zu.

Gleiches gilt im Falle der Aufhebung des Vereins und für den Wegfall des bisherigen Zwecks, ohne daß ein anderer steuerbegünstigter Zweck bestimmt wird. In jedem der drei oben genannten Fälle ist das Vereinsvermögen ausschließlich und unmittelbar für die Zwecke der Förderung der Wissenschaften zu verwenden.

§ 16. Die Mitgliederversammlung kann keine Änderung der Satzung vornehmen, die der Zweckbestimmung des Vereins und des Vereinsvermögens, die Wissenschaft zu fördern, zuwiderläuft.

§ 17. Geschäftsjahr des Vereins ist das Kalenderjahr.

## 9 Die Satzung von 2000

**§ 1 bis 5.** *Unverändert (s. Abschnitt 8).*

**§ 6.** Ordentliches Mitglied kann jede Person werden, sofern sie in der Regel auf den Gebieten, deren Förderung die Vereinigung bezweckt, wissenschaftlich gearbeitet hat. Der Erwerb der Mitgliedschaft erfolgt auf schriftlichen Antrag eines Mitgliedes oder des Bewerbers. Über die Aufnahme beschliesst der Vorstand. Ablehnungen sind in der nächsten Mitgliederversammlung zu bestätigen.

Förderndes Mitglied kann jede juristische Person des In- und Auslandes werden, solange gegen diese Mitgliedschaft nicht vom Vorstand Einspruch erhoben wird.

**§ 7 bis 9.** *Unverändert (s. Abschnitt 8).*

**§ 10.** Ein Mitglied des Vorstandes ist Vorsitzender der Vereinigung, ein weiteres stellvertretender Vorsitzender. Ihre Amtszeiten betragen zwei Jahre. Sie werden von der Mitgliederversammlung gewählt.

In begründeten Ausnahmefällen kann auf Beschluss des Vorstandes die Wahl des Vorsitzenden oder des stellvertretenden Vorsitzenden durch Briefwahl mit einfacher Mehrheit erfolgen.

**§ 11.** Der Vorstand führt die Geschäfte der Vereinigung. Der Vorsitzende und der stellvertretende Vorsitzende sind, jeder für sich alleine, unterschriftsbefugt im Namen des Vorstandes. Sie vertreten die Vereinigung gerichtlich und aussergerichtlich. Im Innenverhältnis darf der stellvertretende Vorsitzende nur vertreten, wenn der Vorsitzende verhindert ist. Der Vorstand fasst seine Beschlüsse mit Mehrheit. Bei Stimmengleichheit entscheidet die Stimme des Vorsitzenden. Der Vorsitzende kann die Entscheidung der Vorstandsmitglieder schriftlich einholen.

Der Vorstand wählt mit einfacher Mehrheit einen Kassenwart aus dem Kreis der Vorstandsmitglieder. Die Verfügung über die Kasse (das Konto) der Vereinigung steht dem Vorstand zu. Unterschriftsberechtigt hierzu ist der Kassenwart. Der Vorstand verteilt einzelne Aufgaben unter sich. Er kann Mitglieder der Vereinigung mit bestimmten Aufgaben befristet betrauen.

**§ 12 bis 17.** *Unverändert (s. Abschnitt 8).*

# Eine kurze Geschichte der Entwicklung der Logik in Münster

Wolfram Pohlers*

Institut für Mathematische Logik und Grundlagenforschung, Fachbereich Mathematik und Informatik, Westfälische Wilhelms-Universität Münster, Einsteinstrasse 62, 48149 Münster, Deutschland
E-Mail: `pohlers@uni-muenster.de`

Der Lehrstuhl für mathematische Logik und Grundlagenforschung ist der älteste Logiklehrstuhl in Deutschland. Deshalb lohnt sich der Versuch eines kurzen Abrisses seiner Enstehungsgeschichte.[1]

Die Geschichte des Lehrstuhls ist untrennbar mit dem Namen Heinrich Scholz verbunden. Heinrich Scholz wurde 1884 in Berlin geboren, studierte in Berlin und Erlangen Philosophie und Theologie, promovierte zum Dr. phil. in Erlangen und wurde 1917 auf einen Lehrstuhl für Religionsphilosophie und systematische Theologie in Breslau und dann 1919 auf einen Lehrstuhl für Philosophie nach Kiel berufen. In der Kieler Bibliothek entdeckte er zufällig ein Exemplar der *Principia Mathematica* von Russell und Whitehead und „bemerkte sofort, dass [er] dort das gefunden hatte, was [er] so lange vergeblich gesucht hatte." Seine langdauernde und hartnäckige Affinität zur mathematischen Logik war damit geboren. Dies motivierte ihn, obwohl bereits Lehrstuhlinhaber, nochmals ein komplettes Studium der Mathematik und Physik zu absolvieren. Unter anderem kam er in Kontakt mit Moritz Schlick, der später den Wiener Kreis mitbegründete, und konnte ihn, allerdings nur für kurze Zeit, nach Kiel holen. In Münster landete Scholz schließlich im Jahre 1928, wieder auf einem Lehrstuhl für Philosophie. In seinem Berufungsschreiben heißt es:

> In Verfolgung der in meinem Auftrage mit Ihnen geführten Verhandlungen sind Sie zum 1. Oktober 1928 in die Philosophische und Naturwissenschaftliche Fakultät der Universität Münster i/W. berufen worden. Ich verleihe Ihnen in dieser Fakultät die durch das Ableben des Professor Brunswig freigewordene planmäßige Professur mit der Verpflichtung, die Philosophie in Vorlesungen und Übungen zu vertreten. Zugleich ernenne ich Sie zum Direktor des Philosophischen Seminars der Universität Münster i./W.

---

*Ich bedanke mich herzlich bei Niko Strobach, nicht nur für die Überlassung des Materials über Heinrich Scholz, sondern auch für seine Bereitschaft, sich auch künftig um seinen Nachlass zu kümmern.

[1] Um keinen falschen Eindruck zu erwecken; die historischen Daten habe ich nicht alle mühsam persönlich recherchiert. Ich habe mich weitgehend auf die Artikel [3] und [2] gestützt.

Der Aufgabenbereich Scholzens wurde dann acht Jahre später auf die Vertretung des Gebiets der logistischen Logik (wie formale Logik damals bezeichnet wurde) und der Grundlagenforschung erweitert. Die Abteilung B des philosophischen Seminars, dessen Leitung mit seiner Berufung an Scholz übertragen wurde, wurde im Jahr 1936 dann in *Logistische Abteilung des Philosophischen Seminars* umbenannt. Aber damit war Heinrich Scholz nicht zufrieden. Offenbar versuchte er hartnäckig seinen Lehrstuhl in die mathematisch-naturwissenschaftliche Abteilung der Philosophischen und Naturwissenschaftlichen Fakultät zu überführen und konnte sich anscheinend mit seinem Begehren auch bei seinen Kollegen durchsetzen. So stellt, sicherlich auf Betreiben Scholzens, die Fakultät noch im Jahr 1938 einen Antrag an das Ministerium in dem es u.a. heißt:

> Das gegenwärtig von Herrn Scholz besetzte Ordinariat wird in dem von Herrn Scholz erbetenen Sinne umgewandelt in ein der mathematisch-naturwissenschaftlichen Abteilung eingeordnetes, an die von Herrn Scholz empfohlene Vorbildung geknüpftes Ordinariat für Philosophie der Mathematik und Naturwissenschaften mit besonderer Berücksichtigung der neuen mathematischen Logik und Grundlagenforschung.

Diesem Antrag wurde noch im gleichen Jahr entsprochen. Aber auch damit war Scholz noch nicht zufrieden, zunächst, weil der Lehrstuhl etatrechtlich nicht selbstständig war. Einem sofort nachgereichten Antrag auf etatrechtliche Verselbstständigung kam das Ministerium jedoch nicht sofort nach. Er wurde zurückgestellt und erst Anfang 1939 genehmigt. Damit hatte Scholz zumindest ein Zwischenziel erreicht. Sein Ordinariat war nun näher an den Interessenbereich herangerückt, der ihm besonders am Herzen lag, und hatte einen eigenen Etat (500 Mark!). Dennoch schien er mit der Bezeichnung seines Ordinariats immer noch nicht wirklich glücklich zu sein. Einem Antrag der Philosophischen und Naturwissenschaftlichen Fakultät entsprechend teilte 1943 der Reichsminister für Wissenschaft, Erziehung und Volksbildung Heinrich Scholz mit, dass

> [der] von Ihnen bekleidete ordentliche Lehrstuhl für Philosophie der Mathematik und Naturwissenschaften mit sofortiger Wirkung in einen solchen für ‚Mathematische Logik und Grundlagenforschung' umgewandelt [wird].

In einem Schreiben vom 10. August 1945 informiert Scholz den Universitätskurator von dieser Umwandlung, wobei er eine „endgültige Überführung in die mathematisch-naturwissenschaftliche Sektion" erwähnt, von der im Schreiben des Ministers eigentlich nicht der Rede war. Schmidt am Busch und Wehmeier vermuten hier einen taktischen Schachzug Scholzens, da mit dem genannten Schreiben die Bitte um die Umwandlung der dem Seminar zugeordneten Hilfskraftstelle in eine planmäßige Assistentenstelle verbunden war. Die Bitte wird u.a. nämlich mit den Worten begründet:

> Das Logistische Seminar ist ein spezifiziertes mathematisches Seminar. Ein mathematisches Seminar kommt ohne eine solche Stelle nicht aus.

Obwohl Scholz bereits da von „[... seinem] 1935 in einer siebenjährigen Vorarbeit ins Leben gerufenen ‚Seminar für mathematische Logik und Grundlagenforschung' " spricht, stellt er erst 1946 in einem Schreiben an den Universitätskurator den Antrag, dass sein Seminar auch so benannt werden sollte. Dem Antrag wurde postwendend entsprochen. Die „Erhebung zum Institut" beantragt Scholz im Jahre 1950. Am 24. Juli 1950 wurde endlich das Scholzsche Seminar mit Erlaß des Kultusministeriums in ein *Institut für Mathematische Logik und Grundlagenforschung* umgewandelt und damit eines der Herzensanliegen Scholzens erfüllt. So heißt das Institut noch heute, das nun nach den diversen Umstrukturierungen der Universität am Fachbereich Mathematik und Informatik angesiedelt ist.

Obwohl Scholz keine großen Spuren in der mathematischen Logik hinterlassen hat—es gibt keinen Satz von Scholz, keine auf Scholz zurückgehenden Begriffsbildungen oder ähnliches—ist er für die Entwicklung der mathematischen Logik in Deutschland von ganz wesentlicher Bedeutung. Durch seine weltweiten Kontakte zu praktisch allen Logikern der Entwicklungsjahre der mathematischen Logik hat er ihr entscheidende Impulse gegeben. Er korrespondierte mit Church, Curry, Kleene, Rosser um nur ein paar Namen zu nennen. Paul Bernays wies ihn bereits auf Gerhard Gentzen hin, der dann auch in Münster vorgetragen hat. Ein für 1952 (!) geplanter Vortrag Turings in Münster scheiterte an diplomatischen Hürden. Es gibt aber ein mit einer persönlichen Widmung versehenes Separatum einer Turing Arbeit in Münster, eine der Kostbarkeiten des Scholz Nachlasses.[2]

Insbesondere kümmerte er sich um den Nachlass Freges, der der Universität von dessen Adoptivsohn Alfred 1935 überlassen wurde. Die Herausgabe des Nachlasses Freges konnte Scholz nicht vollenden. Im Laufe der Arbeit an diesem Nachlass sind jedoch etliche Abschriften entstanden, was sich als Segen herausstellte, da der Nachlass offenbar bei einem Bombenangriff auf die Universitätsbibliothek verbrannt ist.[3]

Scholz hat sich während des dritten Reiches mutig für seine logischen Kollegen in den von Deutschland besetzten Gebieten eingesetzt. Zu Scholzens engeren „logischen" Freunden gehörte Jan Łukasiewicz, der Lehrer

---

[2] Universitäts- und Landesbibliothek Münster, Autographen-Sammlung: Turing, Mediennummer 6-00134596-9, III Tu 1.8-6551.

[3] Kai Wehmeier hat, in der Hoffnung der Frege Nachlass könnte sich unter den ausgelagerten Dokumenten der ULB befinden, eindringliche Versuche unternommen, den Nachlass aufzuspüren. Letztlich haben sich aber alle als vergeblich herausgestellt.

Axiomata für Herrn Dr. Max Bense

(1) 1. Glaubensartikel der Schule von Münster:
Es gibt Kalküle, die durch kein Fingerspitzengefühl
ersetzt werden können. Um diese Kalküle soll man sich so
bemühen, dass man keine Arbeit und Anstrengung scheut, die
man aufwenden muss, um sie effektiv zu beherrschen.

(2) 2. Glaubensartikel der Schule von Münster:
Was durch einen Kalkül beherrscht werden kann, soll man
nicht durch Redeweisen beherrschen wollen, deren Genauigkei
grad in keinem Falle auch nur angenähert heranreicht an
den Genauigkeitsgrad einer Kalkülsprache.

(3) 3. Glaubensartikel der Schule von Münster:
Jedes Handwerk ehrt seinen Meister, wenn es so ausgeübt
wird, dass der Ausübende sich auf eine eindeutige Art von
einem Stümper oder von einem tastenden Lehrling unterschei
det. Es gibt keine philosophische Höhe, die irgend einen
noch so hochstehenden Philosophen berechtigt, auf ein sol
ches substantielles handwerkliches Ki Können herabzusehen

(4) 4. Glaubensartikel der Schule von Münster:
Es gibt ein Fingerspitzengefühl, das durch keinen Kalkül
zu ersetzen ist. Dieses Fingerspitzengefühl ist in jedem
ernst zu nehmenden Falle das Ergebnis einer langen, unver
drossenen Arbeit und Mühe. Es ist genau so hochzuhalten,
wie irgend ein Resultat von dieser Art und auf dieser Stuf

(5) 5. Glaubensartikel der Schule von Münster:
Jeder Forscher bedarf dieses Fingerspitzenkalküls, er sei
auch, wer er sei. Der dezidierteste Kalkülforscher ist
nicht ausgenommen. Er bedarf dieses Fingerspitzengefühls
auf seine Art in einem genau so profunden Sinne wie irgend
ein Forscher, der von ihm verschieden, oder wie irgend ein
Forscher, der beliebig weit entfernt ist davon. Er bedarf
dieses Kalkü Gefühls wie er der Fantasie bedarf, ohne die
in keinem Raum dieser Welt etwas Grosses, Erleuchtendes ge
schaffen worden ist oder geschaffen werden wird. Es ist ein
Zeichen von Barbarei, nicht zu sehen, was jeder muss sehen
können, der ernst genommen werden will. Es ist ein Zeichen
von Barbarei, mit der verflossenen Wiener Schule zu sagen,
dass alles, was in unserer Welt von einer gewissen Kalkül
forschung verschieden ist, in den Bereich der Lyrik fällt
oder in den Abgrund, über welchem geschrieben steht:"Viel
Lärm um Nichts". Dies wird hier so klar und so deutlich ge
sagt, dass niemand sich unterstehen soll, der Schule von
Münster die Seelenblindheit zuzuschieben, durch deren Inter

ABBILDUNG 1. Heinrich Scholz, Axiomata für Herrn Dr. Max Bense, 31. Januar 1942; Universitäts- und Landesbibliothek Münster, Scholz-Nachlaß, Seite 1.

Tarskis, der nach der Besetzung Polens durch Deutschland seine Professur verlor. Scholz erreichte, dass Łukasiewicz 1938 Ehrendoktor der Universität Münster wird und unterstützte später das Ehepaar Łukasiewicz während der deutschen Besatzung nach Kräften. Als die Besetzung Warschaus durch die Rote Armee absehbar wurde, eine ernstliche Bedrohung für das Ehepaar Łukasiewicz, gelingt es ihm, beide nach Münster zu holen. Scholz war auch einer der Verbindungsmänner zwischen Tarski und seiner Familie in Polen. Tarski hat sich brieflich und mit einer Buchwidmung bei Scholz explizit dafür bedankt.

Zusammen mit Hans Hermes und Gisbert Hasenjaeger—beide spätere Gründungsmitglieder unseres Vereins—bildet Scholz die „Schule von Münster", wie sie sich (scherzhaft?)[4] selbst bezeichnen.

Scholz wird im Jahre 1953 emeritiert. Sein Nachfolger wird Hans Hermes, der dann 1966 auf einen neugeschaffenen Lehrstuhl für Mathematische Logik und Grundlagenforschung an die Universität Freiburg wechselt. Gisbert Hasenjaeger erhält 1962 einen Lehrstuhl am neugeschaffenen Seminar für Logik und Grundlagenforschung an der Universität Bonn.[5] Zu seinen Doktoranden gehören Ronald Jensen und Dieter Rödding, der 1966 Nachfolger von Hans Hermes auf dem Münsteraner Lehrstuhl wird. Im Rahmen der allgemeinen Universitätserweiterung wird 1972 auf Betreiben Röddings das Institut für Mathematische Logik und Grundlagenforschung um einen weiteren Lehrstuhl erweitert, auf den 1973 Justus Diller berufen wird.

Dieter Rödding hat sich unter anderem mit endlichen Automaten und abstrakten Maschinen nicht nur theoretisch beschäftigt, sondern diese zum Teil auch technisch realisiert[6], und wurde so zu einem der Vorreiter der Informatik, in der auch die Mehrzahl seiner Schüler ihre Heimat fanden. Zu Röddings Schülern gehören u.a. Helmut Schwichtenberg, der Nachfolger Schüttes in München, Egon Börger, später Professor für Informatik an der Universität Pisa und Thomas Ottmann, später Professor für Informatik in Freiburg. Rödding starb unerwartet im Juni 1984.

Auf den nun frei gewordenen Lehrstuhl wurde ich im Oktober 1985 berufen. Aus endlichen Automaten wurden so unendliche Ordinalzahlen. Diese Berufung, die ja weg von der aufkeimenden Informatik führte, sorgte für einige Aufregung innerhalb einer Gruppe unserer Forschungsgemeinschaft,

---

[4]Es gibt sechs schriftlich festgehaltene Glaubensartikel der Schule von Münster, die vermuten lassen könnten, dass dies vielleicht nicht so ganz ernsthaft gemeint war. Vgl. Abb. 1 & 2.

[5]Vgl. E. Brendel & R. Stuhlmann-Laeisz, Geschichte des Lehrstuhls für Logik und Grundlagenforschung an der Rheinischen Friedrich-Wilhelms-Universität Bonn, in diesem Bande.

[6]Bei der Übernahme des Lehrstuhls fand ich eine Fülle technischen Materials und technischer Geräte vor. Den Großteil davon habe ich zusammen mit Gisbert Hasenjaeger gesichtet und ihm übergeben. Die Geräte wurden später von den Erben Hasenjaegers und Röddings an das Heinz Nixdorf MuseumsForum in Paderborn übergeben [1].

-2-

polation auch die rechtschaffenste Arbeit vor einem gewissen Publikum nach dem erprobten Prinzip des kleinsten Kraftmasses in drei Schritten zum Verschwinden gebracht werden kann. Oder er wundere sich nicht darüber, dass es sich zuträgt auf eine unerwartete Art, dass wir ihn nicht sanfter niedersetzen, als er niederzusetzen zu werden verdient.

(6) 6. Glaubensartikel der Schule von Münster:

Es gibt nichts in der Welt, was einer guten Sache so unüberwindlich schaden kann wie ein Mangel an Geistesgegenwart in irgend einem entscheidenden Augenblick. Ein solcher Mangel muss jedem nachgesagt werden, der sich selbst oder anderen einzureden versucht, dass das ernst zu nehmende Philosophieren von heute oder von morgen ab beschränkt werden könne oder beschränkt werden müsse, auf den engsten Raum der Themen und Fragestellungen, welche die Kalkülforschung bis heute zu bezwingen vermag. Man mache sich klar, was man auf diese Art effektiv tut. Man liefert die wichtigsten Themen und Fragestellungen, die einen meditierenden Menschen bewegen können, einer unkontrollierten schweifenden Jugend oder dem lebenslänglichen Vagantentum aus, wenn man sie auf diese Art loszuwerden versucht. Die Welt ist nicht aus Brei und Mus geschaffen. Niemand darf sich unterstehen, ihre herbe Realität einem persönlichen Wunschgebilde zu opfern, und wenn es die Inseln der Seligen wären, auf die er uns verpflanzen möchte. Es wird und soll und muss neben der philosophischen Grundlagenforschung, für welche die Schule von Münster sich einsetzt, eine zweite philosophische Grundlagenforschung geben, die da einspringt, wo unserer Werkstatt die Grenzen gezogen sind, die sie auf eine ehrliche Art in einem übersehbaren Zeitraum nicht überschreiten wird. Der persönliche Einsatz, den dieses Philosophieren verlangt, ist zu respektieren, so lange es Tag ist und so lange wir einer Welt angehören, die den Landsknechten nicht überantwortet werden darf. Das Einzige, was sich nicht zutragen soll und was wir immer wieder einmal erzwingen werden, wenn es notwendig ist, ist dies, dass diese ganz andere Grundlagenforschung sich nicht in unsere Räume einmischt und dass sie sich hütet vor dem vornehmen Ton, zu dem sie durch nichts berechtigt ist.

Münster i.W., d.31.Jan.1942       Heinrich Scholz

ABBILDUNG 2. Heinrich Scholz, Axiomata für Herrn Dr. Max Bense, 31. Januar 1942; Universitäts- und Landesbibliothek Münster, Scholz-Nachlaß, Seite 2.

da der eigentlich erwartete Nachfolger nicht zum Zuge kam. Selbst habe ich von diesen Aufregungen aber erst dadurch erfahren, dass mir nach meiner Berufung eine kritische Dokumentation des Berufungsverfahrens anonym zugespielt wurde, für die der damalige Vorsitzende der DVMLG, Michael Richter, verantwortlich zeichnete.[7] Innerhalb des Fachbereiches wurde diese Kritik aber entweder nicht wahr oder nicht ernst genommen und ich wurde mit offenen Armen empfangen.

Die Forschungsaktivitäten des Instituts gingen daher jetzt in Richtung Beweistheorie. Bei Justus Diller mit Schwerpunkt Intuitionismus und Funktionalinterpretationen—er hatte bereits in den späten 1970er Jahren zusammen mit Anne Troelstra einen Kontakt zwischen Amsterdam und Münster eingerichtet, auf dessen Basis jährlich ein Logik Kolloquium abwechselnd in Münster und Amsterdam stattfand und der bis etwa 1990 gut funktionierte—bei mir mit Schwerpunkt Ordinalzahlanalysen kombiniert mit abstrakter Rekursionstheorie. Wir haben dabei aber stets darauf geachtet ein gemeinsames Oberseminar zu veranstalten. Meine vier Münsteraner Habilitanden sind heute alle außerhalb Deutschlands tätig. Michael Rathjen als Professor in Leeds, Andreas Weiermann als Professor in Gent, Arnold Beckmann als Professor in Swansea und Gunnar Wilken als Wissenschaftler am *Okinawa Institute for Science and Technology* in Japan. Justus Diller wurde 2001 emeritiert. Sein Nachfolger wurde Ralf Schindler, womit sich einer der Schwerpunkte in Richtung Mengenlehre verlagerte.

Ich selbst wurde im Jahre 2008 pensioniert. Meine Nachfolgerin wurde Katrin Tent mit den Schwerpunkten Modelltheorie und Algebra. Die Schwerpunkte des Instituts haben sich also völlig verlagert. Damit ist das Institut auch näher an das Mathematische Institut gerückt, was sich für die Logik als vorteilhaft erwiesen hat und sicherlich auch im Sinne Scholzens wäre, der ja immer die Nähe zur Mathematik gesucht hat. Heute lehren am Institut für Mathematische Logik und Grundlagenforschung als Professoren Ralf Schindler, Katrin Tent und Martin Hils in Mengenlehre und Modelltheorie. Daneben wirken als Juniorprofessoren Franszizka Jahnke (in der Modelltheorie) und Farmer Schlutzenberg (in der Mengenlehre).

Ob die sechs Glaubensartikel der Scholzschen Münsteraner Schule ernst gemeint waren und so auch noch heute für die mathematische Logik gelten? Dazu möge sich jeder selbst eine Meinung bilden. Im Zusammenhang sind die ersten vier auf Seite 140 des Artikels von Niko Strobach nachzulesen. Abb. 1 & 2 zeigen das ursprüngliche Dokument mit freundlicher Genehmigung der Universitäts- und Landesbibliothek Münster. Persönlich würde ich höchstens den vierten Glaubensartikel akzeptieren.

---

[7]Meine Person spielte in dieser Dokumentation allerdings keine Rolle. Richter, den ich sehr gut kannte, hat es mir gegenüber später allerdings bedauert, sich auf diese Dokumentation eingelassen zu haben.

## Literaturverzeichnis

[1] R. Glaschik, Turing machines in Münster, in: S. B. Cooper, J. van Leeuwen (Hrsgg.), *Alan Turing. His work and impact.* Elsevier 2013, S. 71–77.

[2] H.-C. Schmidt am Busch, K. Wehmeier, „Es ist die einzige Spur, die ich hinterlasse": Dokumente zur Entstehungsgeschichte des Instituts für Mathematische Logik und Grundlagenforschung, in: H.-C. Schmidt am Busch, K. Wehmeier (Hrsgg.), *Heinrich Scholz: Logiker—Philosoph—Theologe.* Mentis, 2005, S. 93–101.

[3] N. Strobach, Heinrich Scholz, Eine Dokumentation, in: R. Schmücker, J. Müller-Salo (Hrsgg.), *Pietät und Weltbezug. Universitätsphilosophie in Münster.* Brill Mentis, Paderborn, 2020, S. 125–158.

# Logik am Mathematischen Institut der Ludwig-Maximilians-Universität München

Wolfram Pohlers[1], Stanley Wainer[2,*]

[1] Institut für Mathematische Logik und Grundlagenforschung, Fachbereich Mathematik und Informatik, Westfälische Wilhelms-Universität Münster, Einsteinstrasse 62, 48149 Münster, Deutschland

[2] 63 Church Street, Addingham, West Yorkshire, LS29 0QS, England
E-Mail: pohlers@uni-muenster.de, pmt6ssw@leeds.ac.uk

Die Geschichte der Logik an der Ludwig-Maximilians-Universität München (LMU) beginnt im Jahre 1966 mit der Berufung Kurt Schüttes auf den neu errichteten Lehrstuhl für Mathematische Logik.

**Kurt Schütte.** Schütte wurde am 18. Oktober 1909 in Salzwedel, Altmark geboren. Er studierte Mathematik in Berlin und Göttingen und promovierte als letzter Doktorand Hilberts 1933 mit einer Arbeit mit dem Titel *Untersuchungen zum Entscheidungsproblem der mathematischen Logik* (Schütte, 1934). Nach Tätigkeiten als Meteorologe im Reichswetterdienst ging er nach Beendigung des Krieges zunächst in den Schuldienst, legte das Assessorexamen ab und wurde nach Hilfskrafttätigkeiten an der Universität Göttingen Assistent von Arnold Schmidt in Marburg. Dort habilitierte er sich im Jahre 1952 und nahm nach Gastprofessuren in Princeton, an der ETH Zürich und an der Pennsylvania State University 1963 den Ruf auf eine Professur an der Universität Kiel an, die er bis zu seiner Berufung nach München innehatte.

Nachdem sich Schütte bis zur Mitte der 1950er Jahre neben der Logik auch noch mit den Grundlagen der Geometrie und Lagerungsproblemen beschäftigt hatte—so gelang ihm gemeinsam mit B. L. van der Waerden die Lösung des sogenannten *Kusszahlenproblems* in drei Dimensionen—, konzentrierte er sich zum Ende der Fünfzigerjahre im Wesentlichen auf Beweistheorie. 1959 erschien in den Grundlehren der Mathematik seine Monographie *Beweistheorie* (Schütte, 1960), in der er unter anderem den von Gerhard Gentzen vorgezeichneten Weg der beweistheoretischen Analyse von Axiomensystemen mit Hilfe der Cantorschen Ordinalzahlen fortführte und perfektionierte.

Schüttes herausragender Beitrag zur Weiterentwicklung des Gentzenschen Programms der ordinalzahlorientierten Beweistheorie ist die systematische Anwendung halbformaler Systeme, d.h. von Kalkülen, die sich zwar

---

*Dieser Artikel ist eine Teilübersetzung des in der Zeitschrift *mathe-lmu.de* der *Carathéodory-Gesellschaft für Förderung der Mathematik in Wirtschaft, Universität und Schule an der LMU München* erschienenen Artikels (Pohlers und Wainer, 2016); die Veröffentlichung erfolgt mit freundlicher Genehmigung der Carathéodory-Gesellschaft.

ABBILDUNG 1. *Links.* Kurt Schütte. *Rechts.* Helmut Schwichtenberg bei der Mitgliederversammlung der Gesellschaft für Mathematische Forschung e.V. im Jahre 2006. (Bilder: Konrad Jacobs / Gerd Fischer. Quelle: Bildarchiv des Mathematischen Forschungsinstituts Oberwolfach.)

an den Gentzenschen Schlussweisenkalkülen orientieren, jedoch Schlussregeln mit unendlich vielen Prämissen zulassen. Beweise in halbformalen Systemen werden durch unendliche fundierte Bäume repräsentiert, deren Tiefe sich in kanonischer Weise durch Cantorsche Ordinalzahlen messen lässt. Auf diese Weise erfuhren die kryptischen Ordinalzahlzuordnungen im originalen Gentzenschen Widerspruchsfreiheitsbeweise für die reine Zahlentheorie eine natürliche Deutung. Eine wesentliche Anwendung erfuhren halbformale Systeme im Beweis der Takeutischen Fundamentalvermutung durch William Tait, Moto Takahasi und Dag Prawitz, die sich alle auf ein semantisches Äquivalent zur Fundamentalvermutung gründen, das Schütte mit Hilfe eines halbformalen Systems entwickeln konnte.

Ein Höhepunkt Schüttes Schaffens war die exakte Bestimmung der Grenze für die prädikativ zu rechtfertigenden Ordinalzahlen, ein Ergebnis, das unabhängig von ihm auch von Solomon Feferman in Stanford erzielt wurde und über das Schütte in einem einstündigen Hauptvortrag auf dem Internationalen Mathematikerkongress 1966 in Moskau berichtete.

1973 wurde er Mitglied der Bayerischen Akademie der Wissenschaften und 1984 korrespondierendes Mitglied der Österreichischen Akademie der Wissenschaften. Er war einer der Begründer der Zeitschrift *Archiv für mathematische Logik und Grundlagenforschung* und gehörte zu den Gründungsmitgliedern der *Deutschen Vereinigung für Mathematische Logik und Grundlagen der Exakten Wissenschaften.* Von 1969 bis 1975 war er Mitorganisator der damals jährlich stattfindenden Logik-Tagungen in Oberwolfach. Er wurde im Jahr 1977 emeritiert und verstarb am 18. August 1998 in München.

Prägend für die Entwicklung der Logik in München waren Schüttes konzise und glasklare Vorlesungen. Er war ein Meister ausgefeilter mathematischer Technik. Von ihm vorgeführte Beweise waren in der Regel optimal und technisch kaum zu verbessern. Seine Vorlesungen zu verfolgen war der reine Genuss. Sie konnten den Hörern eine Ahnung der Befriedigung vermitteln, die man nach einem ästhetisch gelungenen Beweis empfindet. Alle seine Habilitanden und viele seiner Doktoranden wurden erfolgreiche Hochschullehrer. So ist es nicht verwunderlich, dass Kurt Schütte als Begründer einer Schule gilt, die heute als Münchener Schule der Beweistheorie bezeichnet wird, die sich neben der in Stanford entstandenen kalifornischen und der auf Takeuti zurückgehenden japanischen Schule durchaus behaupten kann. Auf der internationalen Tagung zum hundertsten Geburtstag von Gerhard Gentzen 2009 in Leeds wurde der Großteil der Vorträge von Doktor-Söhnen, Doktor-Enkeln oder gar Doktor-Urenkeln von Kurt Schütte gehalten. Von den zweiundzwanzig Autoren des Tagungsbandes *Gentzen's Centenary: The quest for consistency* (Kahle und Rathjen, 2015) gehören allein zehn zu dieser Kategorie. Auch einer der derzeitigen Vizepräsidenten der LMU, Martin Wirsing, ist ein Schütte-Schüler.

Verstärkt wurde die Logik in München zunächst von Privatdozenten, die sich an Schüttes Lehrstuhl habilitierten:

Justus Diller im Jahre 1969, der mit Schütte von Kiel nach München gekommen war. Bis zu seiner Wegberufung nach Münster im Jahre 1973 deckte er die Bereiche Intuitionistische Logik und Funktionalinterpretationen ab.

Horst Osswald im Jahre 1973, der von Hannover nach München gekommen war, kümmerte sich bis zu seiner Pensionierung 2006 um die Bereiche Modelltheorie und insbesondere Nonstandard Analysis und deren Anwendungen.

1978 habilitierten sich gleich drei von Schüttes Assistenten:

Wolfgang Maass, der allerdings nie in München lehrte, sondern zunächst ein Heisenberg Stipendium vorwiegend am *Massachussetts Institute of Technology* wahrnahm, dann *Associate* und später *Full Professor* an der *University of Chicago* wurde und schließlich 1991 einen Ruf auf eine Professur für Theoretische Informatik an der Universität Graz annahm,

Wolfram Pohlers, der 1985 einen Ruf auf einen Lehrstuhl für Mathematische Logik an der Westfälischen Wilhelms-Universität Münster annahm und

Wilfried Buchholz, der als der wohl profilierteste Schüler Schüttes dessen Erbe im Bereich der Ordinalzahlanalysen an der LMU fortführte.

**Helmut Schwichtenberg.** Im Jahre 1942 in Sagan (Niederschlesien) geboren, studierte Helmut Schwichtenberg an der Freien Universität Berlin, schrieb seine Dissertation über subrekursive Hierarchien unter Betreuung von Dieter Rödding und habilitierte sich im Jahre 1974.

In dieser Zeit besuchte er Stanford für ein Jahr und wurde maßgeblich von Georg Kreisel, Solomon Feferman und insbesondere durch die Arbeit von Bill Tait über infinitäre getypte $\lambda$-Terme beeinflußt. Danach war er für vier Jahre in Heidelberg, bevor er einen Ruf auf die Nachfolge Kurt Schüttes auf dem Lehrstuhl für Mathematische Logik an der LMU bekam.

Seit seinem Ruf im Jahre 1978 entwickelte und verstärkte Schwichtenberg, gemeinsam mit seinen drei international bekannten Kollegen, Wilfried Buchholz, Hans-Dieter Donder und Horst Osswald (inzwischen alle pensioniert), die weltweite Forschungsreputation der Logikgruppe an der LMU. Die beweistheoretische Tradition Schüttes bildet den starken Kern der Arbeit dieser Gruppe (man beachte z.B. die wichtigen Bücher Buchholz et al., 1981; Buchholz und Schütte, 1988; Troelstra und Schwichtenberg, 1996), aber im Lichte moderner Forschungsentwicklungen entwickelte sich eine breitere Vision der möglichen Anwendungen der Logik. Dies zeigte sich im daraus resultierenden Interesse von vielen herausragenden Studierenden, die sich für die Logik entschieden. Die angebotenen Vorlesungen zeigten ein besonderes Profil, welches die reinen und angewandten Aspekte der Logik miteinander ins Gleichgewicht brachte, z.B. klassische Beweistheorie und Ordinalzahlanalyse (Buchholz), Mengenlehre und infinitäre Kombinatorik (Donder), Nichtstandardanalysis und ihre Anwendungen in der Stochastik (Osswald) und konstruktive Mathematik und Verwendung von Beweisassistenten (Schwichtenberg).

Die Liste der Promotionen und Habilitationen in Logik aus dieser Zeit ist beeindruckend: Schwichtenberg alleine hatte siebenundzwanzig Doktorandinnen und Doktoranden. Die Liste der Absolventen der Logikgruppe und der vielen internationalen Wissenschaftler, die von der Gruppe so herzlich aufgenommen wurden, enthält viele Namen von Personen, die nun akademische Führungspositionen an Europäischen Universitäten innehaben. Einige Beispiele sind: Ulrich Berger, *Reader* an der *Swansea University*; Ralph Matthes, *Chargé de Recherche* am IRIT in Toulouse; Martin Ruckert, Professor an der Hochschule München; Ulf Schmerl, Professor an der Universität der Bundeswehr München, inzwischen pensioniert); Monika Seisenberger, *Associate Professor* an der *Swansea University* und stellvertretende Fachbereichsleiterin der Informatik; Anton Setzer, *Reader* an der *Swansea University*. Besonders hervorzuheben ist Gerhard Jäger (Professor an der Universität Bern), dessen Promotions- und Habilitationszeit sowohl mit der Ära Schütte als auch der Ära Schwichtenberg überlappte. Seine Berner Forschungsgruppe und die Logikgruppe an der LMU haben viele Jahre lang regelmäßige gemeinsame Treffen abgehalten (die sogenannten *ABM-Workshops*).

Schwichtenbergs nachhaltiges Interesse am Zusammenhang zwischen Beweisen und Berechnungen (dies ist auch der Titel des Werks von Schwichtenberg und Wainer, 2012) wurde durch Kreisels *unwinding programme* beeinflußt: die Idee, daß man aus einem konstruktiven Beweis einer Programmspezifikation unter Verwendung von Normalisierungen und Schnitteliminationen (für die Schwichtenberg eine der führenden Forscherpersönlichkeiten in der Welt ist) ein notwendigerweise korrektes Computerprogramm ablesen kann. Während eines Forschungsfreisemesters an der *Carnegie Mellon University* in den späten 1980er Jahren entwarf Schwichtenberg die Anfänge einer praktischen Implementation des Kreiselschen Programms: dies ist MINLOG, ein System für interaktive Beweisentwicklung und Programmextraktion, welches Schwichtenberg mit seinen Schülerinnen und Schülern im letzten Vierteljahrhundert weiterentwickelte. Das System MINLOG und die mit ihm verbundene Forschung in konstruktiver Mathematik sorgte für internationales Interesse und viele drittmittelfinanzierte Projekte, u.a. das Graduiertenkolleg *Logik in der Informatik* (1997–2006), dessen Sprecher Schwichtenberg war und viele Kooperationsprojekte mit anderen Forschungsgruppen in Europa, Japan und Neuseeland.

Auch innerhalb des Mathematischen Instituts war Schwichtenberg ein wichtiger Einflußnehmer und zweimal Dekan der Fakultät; im Jahre 1986 wurde er zum Mitglied der Bayrischen Akademie der Wissenschaften gewählt und er war von 2001 bis 2008 Mitglied des Wissenschaftlichen Beirats des Mathematischen Forschungsinstituts Oberwolfach. Die vier Münchner Professoren waren und sind Mitglieder von Herausgebergremien der wichtigsten Logikzeitschriften und Buchreihen, sowie regelmäßig in Programmkomitees der bedeutendsten Konferenzen und Sommerschulen des Fachs.

Insgesamt hat die Mathematische Logik an der LMU München die weltweite Entwicklung der Mathematischen Logik wesentlich beeinflusst und kann so auf eine stolze Geschichte zurückblicken.

## Literaturverzeichnis

Buchholz, W., Feferman, S., Pohlers, W., und Sieg, W. (1981). *Iterated inductive definitions and subsystems of analysis: recent proof-theoretical studies*, Band 897 in *Lecture Notes in Mathematics*. Springer-Verlag.

Buchholz, W. und Schütte, K. (1988). *Proof theory of impredicative subsystems of analysis*, Band 2 in *Studies in Proof Theory. Monographs*. Bibliopolis.

Kahle, R. und Rathjen, M., Herausgeber (2015). *Gentzen's centenary. The quest for consistency.* Springer-Verlag.

Pohlers, W. und Wainer, S. (2016). 50 Jahre Logik am Mathematischen Institut der LMU München. *mathe-lmu.de*, 32:10–13.

Schütte, K. (1960). *Beweistheorie*, Band 103 in *Grundlehren der mathematischen Wissenschaften*. Springer-Verlag.

Schütte, K. (1934). *Untersuchungen zum Entscheidungsproblem der mathematischen Logik*. Dissertation, Georg-August-Universität Göttingen.

Schwichtenberg, H. und Wainer, S. S. (2012). *Proofs and computations.* Perspectives in Logic. Cambridge University Press.

Troelstra, A. S. und Schwichtenberg, H. (1996). *Basic proof theory*, Band 43 in *Cambridge Tracts in Theoretical Computer Science*. Cambridge University Press.

# The birth pangs of DLMPS

## Paul van Ulsen

Institute for Logic, Language and Computation, Universiteit van Amsterdam, Postbus 94242, 1090 GE Amsterdam, The Netherlands

**Historical note**

This note was written around 2010 at the request of Wilfrid Hodges, then the President of the *Division for Logic, Methodology and Philosophy of Science* (DLMPS), to explain the history of the institution on its website. It is included in this volume with minimal changes with the permission of the author. In 2015, the DLMPS changed its name to *Division for Logic, Methodology and Philosophy of Science and Technology* (DLMPST) and this also resulted in a name change for its mother organisation the *International Union of History and Philosophy of Science* (IUHPS) which now became the *International Union of History and Philosophy of Science and Technology* (IUHPST). Since this note predates these changes, the old names and acronyms are used.

**Background**

The DLMPS was born in 1955—but its delivery involved complications, many of them surprisingly modern and relevant today. This note provides a brief history.

Several international societies for logic, methodology and the philosophy of science were founded between 1945 and 1950; the most important one was IUPS, the *International Union of Philosophy of Science*.[1] Why was there such a need for large-scale organization after the Second World War? Logical groups existed already in pre-war times in Europe, but they were neither effective lobbies nor democratic associations. And in any case, logic and methodology were not recognized as independent disciplines, and sometimes seen as only semi-serious science. To achieve independence and status, two issues had to be resolved: one ideological, and one practical.

First, the ideology. The prominent Dutch and European logician-philosopher Evert Willem Beth (1908–1964) held that logic and philosophy of science were closely connected; it was bad to separate them. Moreover, only a coalition of the two could achieve independence and university-wide

---

[1] English and French are the official languages of these organizations, and in the early days, the French names and acronyms were generally used; thus, *Union Internationale de Philosophie des Sciences*, abbreviated UIPS. In this text we use the English versions, however, which are more common today. For additional discussion of the *Neuordnung der globalen Wissenschaftswelt* in the 1940s and 1950s, cf. B. Löwe, Grundlagenforschung der exakten Wissenschaften: die DVMLG und die Philosophie, in this volume, § 2.

influence. If logic remains dependent on others, e.g., the mathematicians, then it will be in a marginalized and disdained position,[2] and contacts with philosophy of science and other sciences will be cut off.[3] This was the reason for many people to avoid organizing within the International Mathematical Union (IMU). Only the already existing Association for Symbolic Logic (ASL) was a member of IMU. But on the other hand, nobody wanted to be dependent on the philosophers either, e.g., by joining the *Conseil International de Philosophie et des Sciences Humaines* (CIPSH).[4] Beth was against every philosophical influence, and so was Tarski. The well-known historian of logic Bocheński stated: "We have to fight against both mathematicians and philosophers for its [formal logic's] recognition."[5]

Now the practical politics. An internal association of logicians, not tied to the somewhat America-centric ASL, offered a lot of advantages as a forum, including scientific congresses, journals and books.[6] Then as now, international recognition translated into professorships, influence, and money from national governments.[7] After World War Two, the international community founded large bodies like the United Nations Educational, Scientific and Cultural Organization (UNESCO), which collected many important organizations under its aegis; in particular, the International Council of Scientific Unions (ICSU). Recognition by ICSU, i.e., UNESCO, meant status, national funding, and even some direct United Nations subsidies. No wonder that many attempts were made to reach these flesh-pots of Egypt.

## IUPS and IUHS

In 1949, a Belgian and a Swiss philosopher, Stanislas Dockx (1901–1985) and Ferdinand Gonseth (1890–1975), founded IUPS. They wanted institutional societies as members, and sought ICSU membership. To get the required scientific weight, they needed the ASL, but they hoped to do so at a low cost in voting power. Not every member of the ASL wanted IUPS membership, but Tarski and Church were supporters — they saw it as an improvement of the position of logic all over the world. Church wrote: "I have myself always been interested in an organization of international scope rather than one confined to the United States."[8] Even so, ICSU did not want small unions

---

[2]Tarski to Beth 13 August 1956, 18 December 1960, 30 May 1961; Heyting to Quine (President of the ASL) 17 June 1953; Reichenbach to Châtelet (President of the IUPS) 7 December 1952, Brouwer to Châtelet 20 March 1953.

[3]Beth to Destouches & Feys 8 May 1953, Beth to Rosser (President of the ASL) 21 December 1952.

[4]The DLMPS joined CIPSH in 2011.

[5]Bocheński to Quine (President of the ASL) 15 May 1953.

[6]Bocheński used in 1950, during negotiations with the president and secretary of ICSU, an issue of the JSL to show that logic is real science, and doesn't belong to the humanities alone.

[7]Beth to the DLMPS Committee 13 January 1959, French version c. 1953.

[8]Church to Beth 14 April 1950.

like IUPS as members. Therefore, they asked the International Union of History of Science (IUHS)–already an ICSU member–and IUPS to unite. IUHS, and especially its president Petre Sergescu (1893–1954), did not like that, on the view that sharing means losing. But coerced by ICSU, they opened negotiations with IUPS.

Meanwhile there were changes in IUPS. The autocratic organization got a constitution and became more democratic. The pure logicians took control in 1952.[9] Feys wrote in 1952:

> With the elimination of prof. Gonseth, Dockx and Bayer from the committee, the fundamental difficulties to the adhesion of the ASL to [IUPS] are removed. [...] These three colleagues 1° considered [...] [IUPS] as a tool to secure advantages from UNESCO for the profit of organizations or ideas of their own. 2° [...] they were interested in rather literary forms of 'Philosophy of Science'.[10]

And thus, the ASL became a member of IUPS in 1953.

But still there were negotiations with the unwilling IUHS. ICSU had a good solution: from 1955 on, it stopped the subsidy to IUHS. Another obstacle was removed by the death of Sergescu in 1954. IUHS surrendered, and the result in 1955 was the current International Union of History and Philosophy of Science (IUHPS), as we know it today. It was a union with two divisions, DLMPS for logic, methodology and philosophy of science, and DHS for history of science; each with its own council, and one governing committee for IUHPS as a whole.[11] The IUHPS immediately got entrance to ICSU.

### The early years

Between 1955 and 1958, there was a period of consolidation, though occasional tremors shook the newly-formed coalition. In 1958 there was an international DLMPS congress in Brussels for a fresh constitution. As it happened, the secretary lost all minutes of the meeting. The European members Feys and Beth reconstructed the constitution unofficially in a back room, but passed it off as official afterwards. This relaxed style of management was frequently used by the Europeans, but it proved abhorrent to

---

[9]The new IUPS committee in 1952 consisted of Albert Châtelet (President, 1883–1960), Arend Heyting (1898–1980), Hans Reichenbach (1891–1953), Józef Maria Bocheński (1902–1995), Robert Feys (1889–1961), and Jean-Louis Destouches (1909–1980).

[10]Feys to Rosser (President of the ASL) September 1952. This was also the background for Tarski's attack at a talk during the 1953 IUPS congress on Gonseth, and also for the "war" of already six years with Beth. Concerning Tarski, cf. A. Burdman Feferman and S. Feferman, Alfred Tarski, Life and Logic, pp. 250–252.

[11]The 1955 committee for DLMPS consisted of: Jean Piveteau (President, 1899–1991), Destouches, Beth (also in the central committee of IUHPS); ASL was among the institutional members, and moreover all the national logical societies.

the more legalistic Americans. Tarski and Kleene were furious. Feys, Beth and others had to start all over again, and only after a series of juridical reconstructions, a compromise was reached. DLMPS installed a new committee,[12] and finally there was time for science.

---

[12]Stephen Kleene (President, U.S.A., 1909–1994), Kazimierz Ajdukiewicz (Poland, 1890–1963), Patrick Suppes (U.S.A., 1922–2014), Hans Freudenthal (The Netherlands, 1905–1990).